高等职业教育专业英语系列教材

数控专业英语

主　编　沈延秀　唐利平
副主编　罗　昊　程　文
参　编　郝彦琴　苏　君　熊　毅
主　审　沈言锦

机械工业出版社

本书共分6部分，12个单元，分别介绍了数控加工、数控机床刀具、数控机床结构、数控机床系统、数控机床编程和数控电加工等方面的专业英语知识。本书内容全面、精炼，选材新颖，所有的知识点都围绕数控加工技术展开，难度适中，且每单元的课文后都附有新单词和短语的解释、重点和难点句子的注释、全文翻译。

本书可作为高职高专以及大学本科"数控技术"专业的教材，也可以作为数控技术培训教材，还可以供从事数控相关工作的技术、外贸人员使用。

图书在版编目（CIP）数据

数控专业英语/沈延秀，唐利平主编. —北京：机械工业出版社，2010.9
（2024.7重印）
高等职业教育专业英语系列教材
ISBN 978-7-111-31784-5

Ⅰ.①数… Ⅱ.①沈…②唐… Ⅲ.①数控机床—英语—高等学校：技术学校—教材 Ⅳ.①H31

中国版本图书馆 CIP 数据核字（2010）第 173454 号

机械工业出版社（北京市百万庄大街22号 邮政编码100037）
策划编辑：刘子峰 责任编辑：刘子峰
封面设计：赵颖喆 责任印制：张 博
北京建宏印刷有限公司印刷
2024年7月第1版第7次印刷
184mm×260mm·8.75 印张·209 千字
标准书号：ISBN 978-7-111-31784-5
定价：30.00元

电话服务 网络服务
客服电话：010-88361066 机 工 官 网：www.cmpbook.com
　　　　　010-88379833 机 工 官 博：weibo.com/cmp1952
　　　　　010-68326294 金 书 网：www.golden-book.com
封底无防伪标均为盗版 机工教育服务网：www.cmpedu.com

前 言

本书根据教育部"关于加强高职高专教育教材建设的若干意见"以及高职高专数控专业教学大纲编写而成,从实际出发,力求专业培养的宽口径,具有良好的通用性、实用性和针对性。其次,遵照高等职业教育的应用特性,教材内容力求通俗易懂,便于教学和自学。

本书共分6部分,12个单元,分别介绍了数控加工、数控机床刀具、数控机床结构、数控机床系统、数控机床编程和数控电加工等方面的专业英语知识。本书内容全面、精炼,选材新颖,所有的知识点都围绕数控加工技术展开,难度适中,且每单元的课文后都附有新单词和短语的解释、重点和难点句子的注释、全文翻译。

本书在编排上力求突出实用性,具有以下几个特点:

1. 每个单元的内容在编排上重点突出。各单元的文章后有与之配套的阅读材料,可扩充学生相关领域的知识,满足英文功底较好的读者的需求。

2. 阅读内容贴切实用,选取的单词专业性强,便于学生日后应用。

3. 图文并茂,便于学生理解和学习。

本书由沈延秀和唐利平担任主编,罗昊、程文担任副主编,参加编写的老师还有郝彦琴、苏君、熊毅,株洲职业技术学院的沈言锦作为本套丛书的总主编对全部书稿进行了审阅。

由于编者水平有限,书中难免有不足之处,恳请读者批评指正。

编　者

Contents

前言

Part I Introduction to NC

Unit 1 1
   Text   Computer Numerical Control Manufacturing 1
   Translating Skills   科技英语翻译方法与技巧——科技论文写作知识 5
   Reading Material   History of NC 8

Unit 2 12
   Text   Advantages of NC 12
   Translating Skills   科技英语翻译方法与技巧——分离现象 15
   Reading Material   Machines Using NC 16

Part II CNC Machine Tool

Unit 3 21
   Text   Tooling Systems 21
   Translating Skills   科技英语翻译方法与技巧——如何用英语写个人简历 24
   Reading Material   Cutting Tools 26

Unit 4 29
   Text   Tool Radius Compensation 29
   Translating Skills   科技英语翻译方法与技巧——个人简历范例 31
   Reading Material   Tool Length Offsets and Zero Presets 35

Part III CNC Machine Tool Structure

Unit 5 37
   Text   MCU and CPU 37
   Translating Skills   科技英语翻译方法与技巧——怎样写英文求职信 39
   Reading Material   Machine Movements and Control 41

Unit 6 45
   Text   Types and Parts of Machining Centers 45
   Translating Skills   科技英语翻译方法与技巧——英文求职信范例 47
   Reading Material   The Axis System 49

Part IV CNC Machine Tool System

Unit 7 52

	Text	FANUC System Operation Unit—CRT/MDI Panel	52
	Translating Skills	科技英语翻译方法与技巧——it 的用法	54
	Reading Material	FANUC System Operator's Panel	56

Unit 8 ... 59
    Text   Servo Controls ... 59
    Translating Skills   科技英语翻译方法与技巧——英汉习语的文化差异及翻译方法 ... 61
    Reading Material   FANUC-BESK NC System ... 61

Part V CNC Programming

Unit 9 ... 64
    Text   Programming Concepts ... 64
    Translating Skills   科技英语翻译方法与技巧——如何阅读电子产品的英文说明书 ... 67
    Reading Material   Basic Programming ... 69

Unit 10 ... 72
    Text   G-codes ... 72
    Translating Skills   科技英语翻译方法与技巧——电子产品英文使用说明书范例 ... 74
    Reading Material   M-codes ... 76

Part VI CNC EDM

Unit 11 ... 79
    Text   Electric Discharge Machining ... 79
    Translating Skills   科技英语翻译方法与技巧——长难句的翻译 ... 82
    Reading Material   Laser Beam Machining ... 84

Unit 12 ... 86
    Text   Wire-Cut EDM (1) ... 86
    Translating Skills   科技英语翻译方法与技巧——英语否定式的翻译 ... 90
    Reading Material   Wire-Cut EDM (2) ... 92

Appendix 1   参考译文 ... 96
Appendix 2   Vocabulary ... 122
References ... 131

Part I   Introduction to NC

Unit 1

Text

Computer Numerical Control Manufacturing

Definition of Computer Numerical Control and Its Components

A computer numerical control (CNC) machine is an NC machine with the added feature of an on-board computer, which is often referred to as the machine control unit or MCU. Control units for NC machines are usually hard wired. This means that a machine functions are controlled by the physical electronic elements that are built into the controller.[1] Simple put, numerical control is a method of automatically operating a manufacturing machine based on a code of letters, numbers, and special characters. A complete set of coded instructions for executing an operation is called a program. Such programs are stored in RAM or the random-access memory portion of the computer. They can be played back, edited, and processed by the control. Thus, the machine functions are encoded into the computer at the time of manufacture.

The components found in many CNC systems are shown in Figure 1-1.[2]

Machine control unit: generates, stores and processes CNC programs. The machine control unit also contains the machine motion controller in the form of an executive software program.[3]

NC machine: responds to programmed signals from the machine control unit and manufactures the part.[4]

Types of CNC Equipment

Machining centers are the latest development in CNC technology. These systems come equipped with automatic tool changers with the capability of changing up to 90 or more tools.[5] Many are also fitted with the movable rectangular worktables called pallets. The pallets are used to automatically load and unload work pieces. At a single setup machining centers can perform such operations as milling, drilling, tapping, boring, and so on. Additionally by utilizing indexing heads, some centers are capable of executing these tasks on many different faces of a part and at specified angles. Machining centers save production time and cost by reducing the need for moving a part from one machine to another.

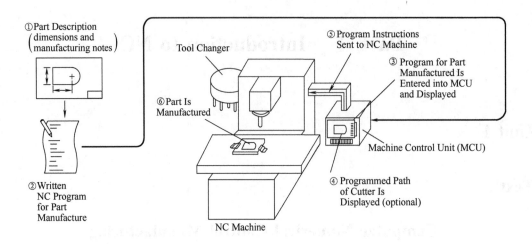

Figure 1-1 Components of Modern CNC System

 Turning centers with increased capacity tool changers are also making strong appearance in modern production shops.[6] These CNC machines are capable of executing many different types of lathe cutting operations simultaneously on rotating part.

 In addition to machining centers and turning centers, CNC technology has also been applied to many other types of manufacturing equipment. Among these are wire-cut electrical discharge machining (wire-cut EDM) machines and laser cutting machines.

 Wire-cut EDM machines utilize a very thin wire as an electrode. The wire is stretched between two rollers and cuts the part like a band-saw.[7] Material is removed by the erosion caused by a spark that moves horizontally with the wire.[8] CNC is used to control horizontal table movements.

 Laser cutting CNC machines utilize an intense beam of focused laser light to cut the part. Material under the laser beam undergoes a rapid rise in temperature and is vaporized. If the beam power is high enough, it will penetrate through the material. Because no mechanical cutting forces are involved, lasers cut parts with a minimum of distortion; they have been very effective in machining slots and drilling holes.[9]

New Words and Phrases

pallet	['pælit]	n.	托盘
milling	['miliŋ]	n.	铣削
drilling	['driliŋ]	n.	钻孔
tapping	['tæpiŋ]	n.	攻螺纹
boring	['bɔːriŋ]	n.	镗孔

lathe　　［leið］　　*n.* 车床　　*vt.* 车削
machining　　［məˈʃiːniŋ］　　*v.* 机械加工
feature　　［ˈfiːtʃə］　　*n.* 特色，功能
wire　　［ˈwaiə］　　*n.* 电线；电报，电信；金属丝，铁丝网　　*vt.* 用金属线捆扎、联结或加固
portion　　［ˈpɔːʃən］　　*n.* 一部分，一份
encode　　［inˈkəud］　　*vt.* 编码，把（电文、情报等）译成电码（或密码）
generate　　［ˈdʒenəˌreit］　　*vt.* 产生，发生
equip　　［iˈkwip］　　*vt.* 装备，配备；训练；准备行装
capability　　［ˌkeipəˈbiliti］　　*n.* 能力，性能；容量，接受力
rectangular　　［rekˈtæŋgjulə］　　*adj.* 长方形的
load　　［ləud］　　*vt. & vi.* 加载
unload　　［ˌʌnˈləud］　　*vi.* 卸载
counter　　［ˈkauntə］　　*adv.* 相反地；反对地
specify　　［ˈspesifai］　　*vt.* 指定；详细说明，列入清单
shop　　［ʃɔp］　　*n.* 商店；工厂，修理厂
simultaneously　　［ˌsiməlˈteiniəsli］　　*adv.* 同时发生地，同步地
electrode　　［iˈlektrəud］　　*n.* 电极；电焊条
discharge　　［disˈtʃɑːdʒ］　　*vt.* 放电；排出，发射　　*n.* 放电量
stretch　　［stretʃ］　　*v.* 伸出；拉紧
roller　　［ˈrəulə］　　*n.* 滚筒，辊子，导轮（用在线切割机床上）
erosion　　［iˈrəuʒən］　　*n.* 腐蚀
spark　　［spɑːk］　　*n.* 火花，火星，电火花
remove　　［riˈmuːv］　　*vt.* 拿走，去掉
intense　　［inˈtens］　　*adj.* 强烈的，剧烈的
focused　　［ˈfəukəst］　　*adj.* 聚焦的
undergo　　［ˌʌndəˈgəu］　　*vt.* 经历，遭受，忍受
vaporize　　［ˈveipəraiz］　　*v.* 蒸发，汽化
penetrate　　［ˈpenitreit］　　*vt. & vi.* 穿透，进入；了解
distortion　　［disˈtɔːʃən］　　*n.* 变形，扭曲
slot　　［slɔt］　　*n.* 狭缝，窄槽
band-saw　　带锯
computer numerical control（CNC）　　计算机数字控制
machine control unit（MCU）　　机床控制单元
electronic element　　电子元器件
random-access memory（RAM）　　随机存储器
machining center　　加工中心

automatic tool changer (ATC)　自动换刀装置
work piece \ workpiece　工件
indexing head　分度头
turning center　车削中心
electrical discharge machining (EDM)　电火花加工
wire-cut electrical discharge machining (wire-cut EDM)　线切割电火花加工
laser cutting machine　激光切割机床
on-board　在船（飞机，车）上，文中指内嵌在机床上
be referred to as...　被称为……

Notes

[1] Control units for NC machines are usually hard wired. This means that a machine functions are controlled by the physical electronic elements that are built into the controller.

NC机床的控制单元通常由硬件构成，也就是说机床的功能是由控制器内置的物理电子元器件控制的。

该句的难点在于wire这个词，wire做名词是"电线、金属丝"的意思，做动词则表示"用金属线捆扎、联结或加固"。在本句中直译应该是：NC机床的控制单元通常是硬连接的。

[2] The components found in many CNC systems are shown in Figure 1-1.

CNC系统中常见的组成部分如图1-1所示。

一般来讲，found不能直接翻译为"找到"，而要根据句子的具体情况翻译为"常见"、"具备"等。

[3] The machine control unit also contains the machine motion controller in the form of executive software.

机床控制单元还包括机床运动控制器，该控制器是一个可执行的软件程序。

in the form of 表示"以……的形式"。

[4] NC machine: responds to programmed signals from the machine control unit and manufactures the part.

NC机床：响应来自机床控制单元的程序信号并对零件进行加工。

[5] These systems come equipped with automatic tool changers with the capability of changing up to 90 or more tools.

这些系统装有自动换刀装置，可换刀具数达90把甚至更多。

[6] Turning centers with increased capacity tool changers are also making strong appearance in modern production shops.

具有大容量刀库的车削中心同样在现代制造厂中占据着显著位置。

[7] The wire is stretched between two rollers and cuts the part like a band-saw.

切割丝在两个导轮之间拉紧,像带锯一样切割工件。

[8] Material is removed by the erosion caused by a spark that moves horizontally with the wire.

切割丝水平移动所产生的点火花会将材料腐蚀并去除。

[9] Because no mechanical cutting forces are involved, lasers cut parts with a minimum of distortion; they have been very effective in machining slots and drilling holes.

由于没有机械切削力,所以激光加工的工件变形非常小,激光切割机对加工窄槽和钻孔非常有效。

Translating Skills

科技英语翻译方法与技巧——科技论文写作知识

A research paper is a form of written academic communication that can be used to give useful information and to share academic ideas with other. Most of the research papers are written for publication in journals or conference proceeding in one's field. Publication is one of the fastest ways for propagating ideas and for professional recognition and advancement. If you have clear idea about the features and styles of academic articles, it will be easier for you to successfully get your paper published in the target journal or accepted by an international conference.

Features of Academic Papers

The first of the features of an academic paper is the content. It is no more and no less than an objective and accurate account of a piece of research you did, either in the humanities, social sciences, natural sciences or applied science. It should not be designed to teach or to provide general background.

The second feature is the style of writing for this purpose. Your paper should contain three ingredients: precise logical structure, clear and simple narration, and the specific style demanded by the journal to which it will be submitted. From the instructions on manuscripts in the "Appendixes A to E", you may get a brief idea of different styles required by different journals.

The third, which is indeed a part of the second, is the system of documenting the sources used in writing the article. At every step in the process of writing, you must take into account the ideas, facts, and opinions you have gained from sources you have consulted.

One of the most convenient features of academic articles is that they are divided into clearly delineated sections. This is helpful because you only have to concentrate on one section at a time. You can thus visualize more or less completely the whole paper while you are working on any part of it. Though papers of the humanities and social sciences do not always have the clearly divided sections, they share some of the common requirements with the scientific papers.

Divisions of Academic Papers

For the average scientific paper the following suggested outline of the divisions of a paper is normally acceptable to, and demanded by, the editors of journals or compilers of conference proceedings:

1) Title of the Paper (subtitle if necessary)

2) Bylines

Name (s) of author (s)

Affiliation of author (s) present and/or permanent address

3) Abstract

The purpose and scope of the paper

The method of study or experiment

A very brief summary of the results, conclusion, and/or recommendations

4) Introduction

A statement of the exact nature of the problem

The background of previous work on this problem done either by the author or others of different approaches

The purpose of this paper

The method by which the problem will be attacked

The primary findings, conclusions and significance of this work

A statement of the organization of the material in the paper

5) Body of Paper

The organization of this main part of the paper is left to the discretion of the author. The information should be presented in some logical sequence, the major points emphasized with suitable illustrations, and the less important ideas subordinated in some appropriate way. This portion of the paper should be styled for the specialist and should not be designed to teach or to provide background for the general reader.

6) Conclusion

Summary and evaluation of the results

Significance and advantages over previous work

Gaps and limitations in the work

Directions for future work and applications

7) Acknowledgements

8) References

学术论文是一种用写作的方式进行学术交流的形式，它可以为他人提供有用的信息并与他人共享学术思想。大多数的学术论文都是为在某一学术领域的期刊发表或为某一领域的会议而写的。出版物是传播学术思想、受到学术界认可和促进学术进步的最快途径之一。如果你对学术文章的特征和类型有很明确的概念，你就更容易使自己的论文在相应杂志上发表或被国际学术会议认可。

学术论文的特征

学术论文的第一特征是内容。客观准确地描述你所进行的一个研究项目，可以是人文科学、社会科学、自然科学，也可以是应用科学。不能把论文写成像教科书或只是提供背景知识。

学术论文的第二特征是为此目的所采用的写作类型。论文应包括三个部分：精确的逻辑结构、清晰简练的叙述以及准备提交的杂志所需要的特定的类型。通过"投稿须知"的"附录 A 到 E"，你可以大概了解各种杂志对不同类型的要求。

学术论文的第三个特征实际上是第二特征的一部分，是用于写文章的资料系统。在写作过程中的每一步，你都要考虑你从查阅的资料中获得的思想、事实和观点。

学术论文的特征中最方便的地方之一是学术论文被清晰地分成几个部分。这一点对你很有帮助，因为你每次只需要考虑一个部分。当你写任一部分时，你总能或多或少地总览全文。尽管人文科学和社会科学的学术论文各部分之间没有明显的界线，它们也部分遵循科学论文的一般要求。

学术论文的分类

一般学术杂志或会议的编辑要求学术论文按如下的大纲分成几个部分：

1）标题（如需要须加副标题）

2）标题下署名之行

作者的名字

联系作者：目前或永久的地址

3）摘要

论文涉及的范围及写作的目的

研究或实验的方法

实验结果、结论的简要概括或评述

4）引言

问题本质的陈述

就此问题，作者或其他人采取不同的方法所做的前期研究的背景情况。

这篇论文的目的

解决问题所要采用的方法

主要的发现，结论和这项工作的意义

介绍论文材料的结构

5）论文主体

论文主要部分的组织方式由作者自己确定，文中信息应遵循某种逻辑顺序，主要论点用适当的例证强调，次重要的观点用适当的方法附于其后。此部分的内容是提供专家同行评阅讨论的，不能写成科教书式的文章或仅为普通读者提供背景知识。

6）结论

小结并评价（实验）结果

（这些工作）相比于前期工作的意义和优点

工作中的局限性和存在的问题

对未来工作和应用的说明

7）致谢，对论文（研究）有帮助的人表示感谢

8）参考文献

Reading Material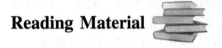

History of NC

Welcome to world of numerical control. Numerical control (NC) has become popular in shops and factories because it helps solve the problem of making manufacturing systems more flexible. In simple terms, a numerical control machine is a machine positioned automatically along a preprogrammed path by means of coded instructions. The key words here are "preprogrammed" and "coded". Someone has to determine what operations the machine is to perform and put that information into a coded form that the NC control unit understands before the machine can do anything. In other words, someone has to program the machine.

Machines may be programmed manually or with the aid of a computer. Manual programming is called manual part programming; programming done by a computer is called computer aided programming (CAP). Sometimes a manual program is entered into the machine's controller via its own keypad. This is known as manual data input (MDI).

Advances in microelectronics and microcomputers have allowed the computer to be used as the control unit on modern numerical control machinery. This computer takes the place of the tape reader found on earlier NC machines. In other words, instead of reading and executing the program directly from punched tape, the program is loaded into and executed from the machine's computer. These machines, known as computer numerical control (CNC) machines, are the NC machines being manufactured today.

In 1947, John Parsons of the Parsons Corporation, began experimenting with the idea of using three-axis curvature data to control machine tool motion for the production of aircraft components. In 1949, Parsons was awarded a U. S. Air Force contract to build what was to become the fist numerical control machine. In 1951, the project was assumed by the Massachusetts Institute of Technology. In 1952, numerical control arrived when MIT demonstrated that simultaneous three-axis movements were possible using a laboratory-built controller and a Cincinnati Hydrotel vertical spindle. By 1955, after further refinements, numerical control became available to industry.

Early NC machines ran off punched cards and tape, with tape becoming the more common medium. Due to time and effort required changing or editing tape, computers were later introduced as aids in programming. Computer involvement came in two forms: computer aided programming languages and direct numerical control (DNC). Computer aided programming languages allowed a part programmer to develop an NC program using a set of universal "pidgin English" commands, which the computer then translated into machine codes and punched into the tape. Direct numerical control (Figure 1-2) involved using a computer as a partial or complete controller of one or more numerical control machines. Although some companies have been reasonably successful at implementing DNC, the expense of computer capability and software and problems associated with coordinating a DNC system renders such systems economically unfeasible for all but the largest companies.

Recently a new type of DNC system called distributive numerical control has been developed (Figure 1-3). It employs a network of computers to coordinate the operation of a number of CNC machines. Ultimately, it may be possible to coordinate an entire factory in this manner. Distributive numerical control solves some of the problems that exist in coordinating a direct numerical control system. There is another type of distributive numerical control that is a spin-off of the system previously explained. In this system, the NC program is transferred in its entirety from a host computer directly to the machine's controller. Alternately, the program can be transferred from a mainframe host computer to a personal computer (PC) on the shop floor where it will be stored until it is needed. The program will then be transferred from the PC to the machine controller.

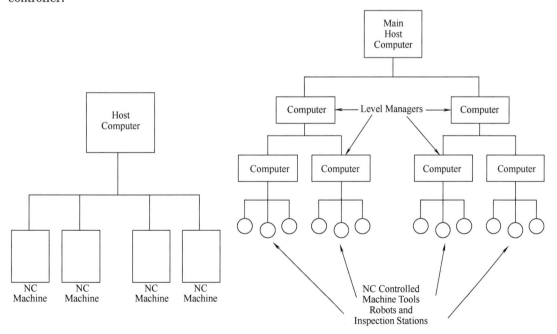

Figure 1-2 Direct Numerical Control   Figure 1-3 Distributive Numerical Control

New Words and Phrases

numerical [nju(ː)'merikəl] *adj.* 数字的
manufacture [ˌmænju'fæktʃə] *v.* 制造，加工 *n.* 制造，制造业；产品
automatically [ˌɔːtə'mætikli] *adv.* 自动地
program ['prəugræm] *vt.* (为……) 编 (制) 程序 *n.* 程序
instruction [in'strʌkʃən] *n.* 指令
preprogram [priː'prəugræm] *v.* 预编程序
code [kəud] *v.* 编码 *n.* 代码，编码
information [ˌinfə'meiʃən] *n.* 信息
manually ['mænjuəli] *adv.* 手动地，人工地
keypad ['kiːpæd] *n.* 键区
data ['deitə] *n.* 数据；参数
input ['input] *n.* 输入 *v.* 输入
microelectronics [ˌmaikrəuiˌlek'trɔniks] *n.* 微电子学
execute ['eksikjuːt] *v.* 执行 *n.* 执行
punch [pʌntʃ] *v.* 冲孔，打孔 *n.* 冲压机，冲床
load [ləud] *v.* 装载，加载
component [kəm'pəunənt] *n.* 部件，零件
technology [tek'nɔlədʒi] *n.* 工艺，技术
spindle ['spindl] *n.* 主轴
medium ['miːdjəm] *n.* 介质，媒介；方法
command [kə'mɑːnd] *n.* 命令 *v.* 命令，指挥
translate [træns'leit] *v.* 转变为；翻译
controller [kən'trəulə] *n.* 控制器
software ['sɔftwɛə] *n.* 软件
coordinate [kəu'ɔːdinit] *v.* 调整；协调 *n.* 坐标
network ['netwəːk] *n.* 网络
transfer [træns'fəː] *v.* 传递；改变 *n.* 传递；转移
mainframe ['meinfreim] *n.* 主机，大型机
store [stɔː] *v.* 存储，储藏 *n.* 储备
numerical control (NC) 数字控制 (数控)
control unit 控制装置，控制单元
manual programming 手工编程
computer aided programming (CAP) 计算机辅助编程
manual data input (MDI) 手动数据输入

tape reader　读带机
punched tape　穿孔带
machine tool　机床
punched card　穿孔卡
direct numerical control (DNC)　直接数字控制
distributive numerical control　分布式数字控制
host computer　主机
personal computer (PC)　个人计算机

Unit 2

Text

Advantages of NC

Recent studies show that of the amount of time an average part spent in a shop, only a fraction of that was actually spent in the machining process. Let us assume that a part spent 50 hours from the time it arrived at a plant as a rough casting or bar stock to the time it was a finished product. [1] During this time, it would be on the machine for only 2.5 hours and be cut for only 0.75 hour. The rest of the time would be spent on waiting, moving, setting up, loading, unloading, inspecting the part, setting speeds and feeds and changing cutting tools.

NC reduces the amount of non-chip-producing time by selecting speeds and feeds, making rapid moves between surfaces to be cut, using automatic fixtures, automatic tool changing, controlling the coolant, in-process gagging, and loading and unloading the part. [2] These factors, plus the fact that it is no longer necessary to train machine operators, have resulted in considerable savings throughout the entire manufacturing process and caused tremendous growth in the use of NC. [3] Some of the major advantages of NC are as follows:

1) There is automatic or semiautomatic operation of machine tools. The degree of automation can be selected as required.

2) Flexible manufacturing of parts is much easier. Only the tape needs changing to produce another part.

3) Storage space is reduced. Simple work-holding fixtures are generally used, reducing the number of jigs or fixtures which must be stored.

4) Small part lots can be run economically. Often a single part can be produced more quickly and better by NC.

5) Nonproductive time is reduced. More time is spent on machining the part, and less time is spent on moving and waiting.

6) Tooling costs are reduced. In most cases complex jigs and fixtures are not required.

7) Inspection and assembly costs are lower. The quality of the product is improved, reducing the need for inspection and ensuring that parts fit as required.

8) Machine utilization time is increased. There is less time that a machine tool is idle because work-piece and tool changes are rapid and automatic.

9) Complex forms can easily be machined. The new control unit features and programming capabilities make the machining of contours and complex forms very easy.

10) Parts inventory is reduced. Parts can be made as required from the information on the

punched tape.

Since the first industrial revolution, about 200 years ago, NC has had a significant effect on the industrial world. The developments in the computer and NC have extended a person's mind and muscle. The NC unit takes symbolic input and changes it to useful output, expanding a person's concepts into creative and productive results. NC technology has made such rapid advances that it is being used in almost every area of manufacturing, such as machining, welding, press-working, and assembly.

If industry's planning and logic are good, the second industrial revolution will have as much or more effect on society as the first industrial revolution had. As we progress through the various stages of NC, it is the entire manufacturing process which must be kept in mind.

Computer aided manufacturing (CAM) and computer integrated machining (CIM) are certainly where the future of manufacturing lies, and considering the developments of the past, it will not be too far in the future the automated factory is reality.[4]

Developed originally for use in aerospace, NC is enjoying widespread acceptance in manufacturing. The use of CNC machines continues to increase, becoming visible in most metalworking and manufacturing industries. Aerospace, defense contract, automotive, electronic, appliance, and tooling industries all employ numerical control machinery. Advances in microelectronics have lowered the cost of acquiring CNC equipment. It is not unusual to find CNC machinery in contract tool, die, and mold-making shops. With advent of low cost OEM (original equipment manufacturer) and retrofit CNC vertical milling machines, even shops specializing in one-of-a-kind prototype work are using CNCs.

Although numerical control machines traditionally have machine tools, bending, forming, stamping, and inspection machines have also been produced as numerical control systems.

New Words and Phrases

load　　　　[ləud]　　v. 装载
inspect　　　[in'spekt]　　v. 检查
feed　　　　[fiːd]　　n. 进给
chip　　　　[tʃip]　　n. 碎片，碎屑　v. 切成碎片
fixture　　　['fikstʃə]　　n. 夹具
coolant　　　['kuːlənt]　　n. 切削液
process　　　[prə'ses]　　n. 过程，步骤　v. 加工，处理
gage　　　　[geidʒ]　　v. 验，校准
semiautomatic　　['semiˌɔːtə'mætik]　　adj. 半自动的
automation　　[ˌɔːtə'meiʃən]　　n. 自动化；自动操作
storage　　　['stɔridʒ]　　n. 贮藏，存储

jig　　　[dʒig]　　n. 夹具
assembly　　[ə'sembli]　　n. 装配，组装
contour　　['kɔntuə]　　n. 轮廓
output　　['autput]　　n. 输出
welding　　['weldiŋ]　　n. 焊接
logic　　['lɔdʒik]　　n. 逻辑；逻辑性
integrate　　['intigreit]　　v. 使一体化，集成
metalworking　　['metəl,wə:kiŋ]　　n. 金属加工术
aerospace　　['ɛərə,speis]　　n. 航空；宇宙空间
automotive　　[ɔ:tə'məutiv]　　adj. 汽车的
tooling　　['tu:liŋ]　　n. 加工；刀具，工具
equipment　　[i'kwipmənt]　　n. 装置，设备，装备
die　　[dai]　　n. 模具，冲模
retrofit　　['retrə,fit]　　n. 改型（装，进）；（式样）翻新
bending　　['bendiŋ]　　n. 弯曲
forming　　['fɔ:miŋ]　　n.（成形）加工
stamping　　['stæmpiŋ]　　n. 冲压
work-holding　　n. 工件夹紧
machining process　　加工过程
rough casting　　铸造毛坯
bar stock　　棒料
finished product　　成品
non-chip-producing time　　非切削时间
tool changing (tool change)　　换刀
in-process gagging　　在线检测
tooling cost　　刀具加工成本
computer aided manufacturing (CAM)　　计算机辅助制造
computer integrated machining (CIM)　　计算机集成加工
tooling industry　　刀具业
prototype work　　标准工件
mold making　　模具制作

Notes

[1] Let us assume that a part spent 50 hours from the time it arrived at a plant as a rough casting or bar stock to the time it was a finished produce.

我们假定把一个零件从运达工厂时的毛坯或棒料加工为成品需要 50 小时。

句中 that 引导的是宾语从句；介词短语 from the time... to the time... 中，time 后分别是省略 that 的定语从句。

[2] NC reduces the amount of non-chip-producing time by selecting speeds and feeds, making rapid moves between surfaces to be cut, using automatic fixtures automatic tool changing, controlling the cool-ant, in-process gagging, and loading and unloading the part.

通过转速与进给量的设定、刀具在待切削表面间的快速移动、自动夹具的运用、刀具的自动切换、切削液的控制、在线检测及零件的自动装卸，NC 缩短了非切削加工时间。

此句虽长，但结构简单：NC reduces... by...。by 表示方法与手段。

[3] These factors, plus the fact that it is no longer necessary to train machine operators, have resulted in considerable savings throughout the entire manufacturing process and caused tremendous growth in the use of NC.

上述因素，加之无需再培训机床操作工，使整个加工过程耗时大量减少，促进了数控的推广应用。

句子主语是 These factors，谓语有两个：have resulted in 与（have）caused。the fact 后由 that 引出同位从句。

[4] Computer aided manufacturing (CAM) and computer integrated machining (CIM) are certainly where the future of the manufacturing lies, and considering the developments of the past, it will not be too far in the future before the automated factory is a reality.

回顾过去的发展历史，可以肯定，未来的制造业属于计算机辅助制造（CAM）与计算机集成加工（CIM），自动化工厂在不远的未来将成为现实。

现在分词短语 considering the developments of the past 作状语。

Translating Skills

科技英语翻译方法与技巧——分离现象

在一般情况下，句中的某些成分应当放在一起，如主语和谓语，动词和宾语等，但在科技英语中我们常可以看到这些成分被隔离开了，这种情况叫分离现象。起隔离作用的主要成分有：

1) 各种短语，如介词短语、分词短语、不定式短语等。
2) 各种从句。
3) 句子附加成分，如插入语、同位语、独立成分。

分析这种隔离现象可以更好地帮助我们理解整个句子的原意。常见的分离现象有以下几种：

1. 主谓分离

The force that pushes you toward the front of the bus when it stops is the inertia of your body.

汽车停止时，推你往前倾的力就是你的身体的惯性。

The keyboard, the most commonly used input device, is introduced in Unit 6.

在 6 单元中介绍键盘——最常用的输入设备。

2. 动宾分离

有时作状语的介词短语等放在动词之后，用来修饰该动词，而把动词描述动作的对象——宾语隔开了。例如：

But naysayers point out that tablet-style computers have been tested with flaws that vendors now claim to have overcome.

但是反对者提出，平板风格的计算机以前就实验过有缺陷，虽然供应商声称已经克服了缺陷。

Overcome 的宾语是 that，或者说是 flaws 。

3. 复合谓语本身的分离

在复合谓语之间插入含有状语意义的介词短语和状语从句，使复合谓语本身产生分离现象。例如：

Many forms of motion are highly complex, but they may in all cases be considered as being made up of translations and rotations.

很多运动的形式是非常复杂的，但在所有情况下，都可以认为是由平移和转动合成的。

复合谓语 may be considered 之间插入 in all cases 造成分离现象。

4. 定语和被修饰名词的分离

Insulation is a material that offers a high resistance to the passage through it of an electric current.

绝缘体是一种对通过电流产生高阻力的材料。

of an electric current 是修饰 passage 的定语，被 through it 分隔开。

5. 某些词与所要求介词的分离

The electric resistance of a wire is the ratio of the potential between its two ends to the current in the wire.

导线的电阻等于该导线两端之间的电位差与导线中电流的比值。

the ratio of A to B 表示"A 与 B 之比"，这里被 between its two ends 隔开了。

Reading Material

Machines Using NC

Early machine tools were designed so that the operator was standing in front of the machine while operating the controls. This design is no longer necessary, since in NC the operator no longer controls the machine tool movements. On conventional machine tools, only about 20 percent of the time was spent removing material. With the addition of electronic controls, actual time spent removing metal has increased to 80 percent and even higher. It has also reduced the amount of time required to bring the cutting tool into each machining position.

In the past, machine tools were kept as simple as possible in order to keep their costs down. Because of the ever-rising cost of labor, better machine tools, complete with electronic controls, were developed so that industry could produce more and better goods at prices which were competitive with those of offshore industries.

NC is being used on all types of machine tools, from the simplest to the most complex. The most common machine tools are the single-spindle drilling machine, engine lathe, milling machine, turning center and machining center.

Single-spindle Drilling Machine

One of the simplest numerically controlled machine tools is the single-spindle drilling machine. Most drilling machines are programmed on three axes:

a. The X-axis controls the table movement to the right and left.

b. The Y-axis controls the table movement toward or away from the column.

c. The Z-axis controls the table movement of the spindle to drill holes to depth.

Engine Lathe

The engine lathe, one of the most productive machine tools, has always been a very efficient means of producing round parts (Figure 2-1). Most lathes are programmed on two axes:

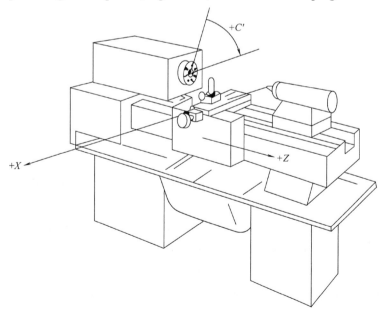

Figure 2-1　The Engine Lathe Cutting Tool Moves Only on the X and Z Axis

a. The X-axis controls the cross motion (in or out) of the cutting tool.

b. The Z-axis controls the carriage travel toward or away from the headstock.

Milling Machine

The milling machine has always been one of the most versatile machine tools used in industry (Figure 2-2). Operations such as milling, contouring, gear cutting, drilling, boring, and reaming are only a few of the many operations which can be performed on a milling machine. The milling machine can be programmed on three axes:

a. The X-axis controls the table movement left or right.

b. The Y-axis controls the table movement toward or away from the column.

c. The Z-axis controls the vertical (up or down) movement of knee or spindle.

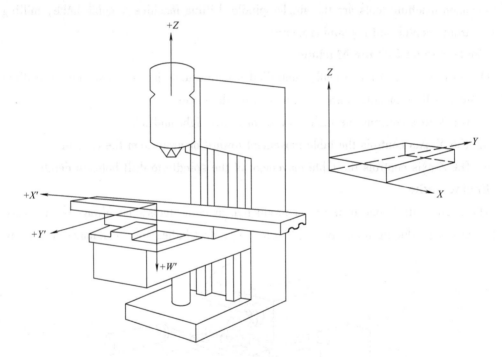

Figure 2-2 The Vertical Knee and Column Milling Machine

Turning Center

Turning Centers were developed in the mid-1960s after studies showed that about 40 percent of all metal cutting operations were performed on lathes. These numerically controlled machines are capable of greater accuracy and higher production rates than were possible on the engine lathe. The basic turning center operates on only two axes:

a. The X-axis controls the cross motion of the turret head.

b. The Z-axis controls the lengthwise travel (toward or away from the headstock) of the turret head.

Machining Center

Machining centers were developed in the 1960s so that a part did not have to be moved from machine to machine in order to perform various operations. These machines greatly increased production rates because more operations could be performed on a workpiece in one setup. There are two main types of machining centers, the horizontal and the vertical spindle types.

a. The horizontal spindle machining center operates on three axes:
(a) The X-axis controls the table movement left or right.
(b) The Y-axis controls the vertical movement (up or down) of the spindle.
(c) The Z-axis controls the horizontal movement (in or out) of the spindle.

b. The vertical spindle machining center operates on three axes:
(a) The X-axis controls the table movement left or right.
(b) The Y-axis controls the table movement toward or away from the column.
(c) The Z-axis controls the vertical movement (up or down) of the spindle.

New Words and Phrases

operator ['ɔpəreitə] n. （机器、设备等的）操作员
operate ['ɔpəreit] v. 运转；操作
electronic [ilek'trɔnik] adj. 电子的；电子器件的
drill [dril] n. 钻头，钻床，钻机
lathe [leið] n. 车床
milling ['miliŋ] n. 铣削
turning ['tə:niŋ] n. 车削
table ['teibl] n. 工作台
column ['kɔləm] n. 立柱
part [pɑ:t] n. 零件，工件
carriage ['kæridʒ] n. （机床的）滑板；刀架
headstock ['hedstɔk] n. 主轴箱
contouring ['kɔntuəriŋ] n. 成形加工
bore [bɔ:, bɔ:(r)] v. 镗（穿、扩、钻）孔
boring ['bɔ:riŋ] n. 镗孔；镗削加工
ream [ri:m] v. 铰孔
reaming ['ri:miŋ] n. 铰孔
knee [ni:] n. 升降台
accuracy ['ækjurəsi] n. 精确性，准确性，精度
workpiece ['wə:k,pi:s] n. 工件
setup ['setʌp] n. 安装；设备；机构

horizontal   [ˌhɔriˈzɔntəl]   *adj.* 水平的
vertical   [ˈvəːtikəl]   *adj.* 垂直的，直立的
cutting tool   刀具
drilling machine   钻床
turning center   车削中心
machining center   加工中心
engine lathe   卧式车床
cross motion   横向运动
gear cutting   齿轮加工
metal cutting   金属切削
production rate   生产率
turret head   转塔头
lengthwise travel   纵向运动

Part II CNC Machine Tool

Unit 3

Text

Tooling Systems

The machining center, a multifunction machine tool, uses a wide variety of cutting tools such as drills, taps, reamers, end mills, boring tools, etc, to perform various machining operations on a workpiece. To be inserted into the machine quickly and accurately, all these cutting tools must have the same taper shank toolholders to suit the machine spindle. The most common taper used in NC machining center spindles is the No. 50 taper, which is a self-releasing taper. The toolholder must also have a flange or collar, for the tool-change arm to grab, and a stud, tapped hole, or some other device for holding the tool securely in the spindle by a power drawbar or other holding mechanism.

When one is preparing for a machining sequence, the tool assembly drawing is used to select all the cutting tools required to machine the part. Each cutting tool is then assembled off-line in a suitable toolholder and preset to the correct length. Once all the cutting tools are assembled and preset, they are loaded into specific pocket locations in the machine's tool-storage magazine where they are automatically selected as required by the part program. [1]

Tool Identification

NC machine tools use a variety of methods to identify the various cutting tools which are used for machining operations. The most common methods of identifying tools are:

1. Tools Pocket Locations

Tools for early machining centers were assigned a specific pocket location in the tool-storage magazine, and each tool was called up for use by the part program.

2. Coded Rings on Toolholders

A special interchange device reader was used to identify some tools by special coded rings on the toolholder.

3. Tool Assembly Number

Most modern MCUs have a tool identification feature which allows the part program to recall a tool from the tool-storage magazine pocket by using a five-to eight-digit tool assembly number.

Each tool assembly number may be assigned a specific pocket in the tool-storage magazine by the tool data tape, by the operator using the MCU, or by a remote tool management console.

Tool Management Program

It is very important, to achieve the best productivity from any machine tool, to have a sound program which covers all aspects of cutting tools. [2] The best NC machine tool can not come anywhere close to its productivity potential unless the best cutting tools for each operation are available for use when they are required. The tool-management program must include such things as tool design, standard coding system, purchasing, good tooling practices, part programming which is most effective, and the best use of cutting tools on the machine.

A good cutting tool policy must include the following:

1. Standard Policy

A standard policy regarding cutting tools must be established.

Every one should clearly understand the policy: the tool engineer, the part programmer, the supervisory staff, the setup person, and the machine tool operator.

The role that each person has in selecting the proper cutting tools must be clearly defined.

2. Cutting Tool Dimensional Standards

All cutting tools purchased or specially made must conform to established cutting tool dimensional standards.

When it is necessary to recondition cutting tools, they should be ground to the NC standard.

The part programmer must use the cutting tool standards for programming purposes.

3. Rigid Cutting Tools

Always select the shortest cutting tool possible for each job to ensure location accuracy and rigidity.

Cutting tool holders should be of one-piece construction to provide rigidity.

4. Tool Preparation

There must be a rigid policy on tool setting, compensation, and regrinding which is understood by everyone concerned.

Clearly define who has the responsibility for each so that there is no conflict or misunderstanding.

5. Indexable Insert Tools

Use cemented carbides insert-type tooling wherever possible because of their wear resistance, higher productivity, and dimensional accuracy.

Borazon (CBN) inserts should be used on hard ferrous metals where cemented carbides are not satisfactory.

Synthetic (polycrystalline) diamond inserts should be used for machining nonferrous materials.

The success or failure of a tool-management program depends in large part on the part programmer. To be most effective, the programmer must have a thorough knowledge of machining practices and procedures and the type of cutting tool required for each operation. Most modern CNC units have standard or optional features or programs available to make any tool-management policy more effective.

New Words and Phrases

multifunction   [ˌmʌltiˈfʌŋkʃən]   n. 多功能
toolholder   [ˈtuːlˌhəuldə(r)]   n. 刀柄
flange   [flændʒ]   n. 凸缘，法兰
collar   [ˈkɔlə]   n. 套环，卡圈，安装环
drawbar   [ˈdrɔːˌbɑː]   n. 拉杆
console   [kənˈsəul]   n. 控制台
regrind   [riˈgraind]   v. 重磨
ferrous   [ˈferəs]   adj. 铁的，含铁的
synthetic   [sinˈθetic]   adj. 合成的
taper   [ˈteipə]   n. 锥度
stud   [stʌd]   n. 双头螺栓；销子；拉钉
hard ferrous metal   硬铁合金
nonferrous material   非铁材料
coded ring   编码环
tool-storage   刀具存放
dimensional standard   尺寸标准

Notes

[1] Once all the cutting tools are assembled and preset, they are loaded into specific pocket locations in the machine's tool-storage magazine where they are automatically selected as required by part program.

一旦所有刀具装设就绪，便可装入机床刀库的指定刀位，以便根据零件程序要求自动选刀。

[2] It is very important, to achieve the best productivity from any machine tool, to have a sound program which covers all aspects of cutting tools.

为了达到最大生产能力，给机床配置功能齐全的刀具管理程序就显得十分重要。

句中 it 为形式主语，不定式 to have a sound program 是主句的真实主语；which 引出的定语从句修饰 program。

Translating Skills

科技英语翻译方法与技巧——如何用英语写个人简历

个人简历能体现一个人的综合实力及英语水平。一篇好的英语简历会给招聘方留下良好印象,从而为获得理想的工作奠定基础。

一、英语简历的写作特点

1. 突出重点

1) 充分体现你的优点与长处,尤其是招聘方要求的知识、经验与技术。

2) 如实列出与招聘相关的业绩和成果。

2. 语言简洁

1) 通常要尽量省略主语"I",因为你的名字已经出现在个人资料栏目中。

2) 省略谓语动词 be。

例如:I was born on January 6, 1984. 应改为 Born: January 6, 1984.

3) 尽量用简单句。例如:As I have been a typist for two years, I can type very quickly and accurately. 应改为 Having been a typist for two years, I can type very quickly and accurately.

4) 使用通用的缩略词。例如:表示身高的 centimeter(厘米),可略为 cm;表示体重的 kilogram(千克),可略为 kg。

3. 篇幅适宜(16 开纸一页即可)

4. 避免差错

在书写简历时必须避免下列错误:1) 语法错误和单词拼写错误;2) 经历不完整;3) 没有写明以前的工作业绩;4) 版面设计混乱。

二、英语简历的构成

1. 个人资料(Personal Data)

1) Name(姓名),例如:Xiaohong Yu(余小红)

2) Address(通信地址),例如:Apt. 308, No. 12, Shanghai Road, Nangang, Henan

3) Postal Code(邮政编码)

4) Phone Number(电话号码)

5) Sex(性别):Male(男)或 Female(女)

6) Date of Birth(出生日期)

7) Birthplace(出生地点),例如:Nanyang

8) Nationality(国籍或民族),例如:P. R. C(中国)或 the Han(汉族)

9) Height(身高),例如:170 cm(170 厘米)或 1.70 m(1 米 70)

10) Weight(体重),例如:60 kg(60 公斤)

11) Health Condition(健康状况),例如:Excellent(极佳)或 Very Good(很好)

12）Hobbies（业余爱好），例如：Play Basketball（打篮球）或 Dancing（跳舞）

13）Number of Identification Card（身份证号码）

2. 应聘职位（Job Objective）

若将个人简历递交人才交流中心、劳务市场或劳务中介机构，务必写出自己所希望获得的职位，如 Assistant Manager, Secretary 等。

3. 学历（Education）

填写学历时，应从最高学历开始，一直向前推移。例如：

September 1999 ～ July 2003（1999 年 9 月 ～ 2003 年 7 月）

Henan Polytechnic Institute（河南工业职业技术学院）

September 1993 ～ July 1999（1993 年 9 月 ～ 1999 年 7 月）

Nanyang No. 29 Middle School（南阳第 29 中学）

September 1987 ～ July 1993（1987 年 9 月 ～ 1993 年 7 月）

Nanyang Shanghai Road Primary School（南阳上海路小学）

为使招聘单位更好地了解你的知识、智力等情况，可将所修课程中与应聘工作有关的学科成绩列出。例如：

Curriculum Grades（所修课程成绩）

English：98（英语：98）

Higher Mathematics：92（高等数学：92）

General Physics Laboratory：B（普通物理实验：良）

4. 社会实践（Social Practice）

主要是指在校内外参与业余工作（Part-time Jobs）、暑假打工（Summer Jobs）和毕业实习（Graduation Practice）等。例如：

From 1999 to 2001, served as governess helping a middle school student with his maths and physics, twice a week.（1999 年至 2001 年期间，当过家庭女教师，辅导了一名中学生的教学和物理，每周两次。）

Graduation Practice：Work at Nanjing SHARP Electronics Corporation Ltd. from February to June 2003.（毕业实习：2003 年 2 月至 6 月在南京夏普电子有限公司工作。）

5. 业余爱好（Spare Hobbies）

1）体育活动（Sports），例如：

Won championship in women eight-hundred-meter race at the school sports meet in 1998.（在 1998 年的校运会上获得女子 800 米冠军。）

2）文娱活动（Entertainment），例如：

Won the first place in the school chess match in 1997.（1997 年获全校象棋第一名。）

Member of the school choir.（校合唱队成员。）

6. 奖励（Rewards）

例如：Elected a Three Goods Student in 1998.（1998 年被评为"三好学生"。）

Won the title of Excellent Leader Council in 1998.（1998 年获得"校学生会优秀干部"

称号。)

7. 技术资格和特别技能（Technical Qualifications and Special Skills）

1）技术资格（Technical Qualifications），例如：

Took out a driving license in 1999. （1999年领取驾驶执照。）

Got 6th class electrician certificate in 1999. （1999年获得六级电工证书。）

2）特别技能（Special Skills），例如：

English：College English Test-Band Four（英语：大学英语四级）

Computer Language：BASIC，PASCAL（计算机语言：BASIC，PASCAL）

Typing Proficiency：60 wpm；Shorthand：90 wpm（打字熟练程度：每分钟60个单词；速记：每分钟90个单词）

8. 科研成果（Scientific Research Achievements）

1）发表或出版的作品、论文、论著等（Publications），写明作品的题目、发表或出版的时间、发表或出版作品的报刊杂志或出版社的名称。

2）创造发明（Invention），写出发明物的名称、时间和地点。

3）技术革新（Technical Innovations），写清革新项目、时间、地点与经济效益。

9. 证明人（References）

1）证明你的学业情况及性格——校长、专业主任、任课老师等。

2）证明你的工作能力、业务水平和个性——原工作单位的领导。

证明人应包括其姓名、职位、单位电话和地址，这样招聘单位就会认为你的简历是真实的，而且是经得起查询的。

另外，在寄个人简历时还要附上毕业证书、技术资格证的复印件。

Reading Material

Cutting Tools

The selection of the proper cutting tools for each operation on machining center is essential to producing an accurate part. Generally there is not enough thought and planning going into the selection of cutting tools for each particular job. The NC programmer must have a thorough knowledge of cutting tools and their applications in order to properly program any part.

Machining centers use a variety of cutting tools to perform various machining operations. These tools may be conventional high-speed steel, cemented carbide inserts, CBN (cubic boron nitride) inserts, or polycrystalline diamond insert tools. Some of the tools used are end mills, drills, taps, reamers, boring tools, etc.

Studies show that machining center time consists of 20 percent milling, 10 percent boring, and 70 percent hole-making in an average machine cycle. On conventional milling machines, the cutting tool cuts approximately 20 percent of the time, while on machining centers the cutting time can be as high as 75 percent. The end result is that there is a larger consumption of disposable tools due to decreased tool life through increased tool usage.

End Mills

End mills and shell end mills are widely used in machining centers. They are capable of performing a variety of machining operations such as face, pocket, and contour milling; spot facing; counter boring; and roughing and finishing of holes using circular interpolation.

Drills

Conventional as well as special drills are used to produce holes. Always choose the shortest drill that will produce a hole of the required depth. As drill diameter and length increase, so does the error in hole size and location. Stub drills are recommended drilling on machining centers.

Center Drills

Center drills are used to provide an accurate hole location for the drill which is to follow. The disadvantage of using center drills is that the small pilot drill can break easily unless care is used. An alternative to the center drill is the spotting tool, which has a 90° included angle and is widely used for spotting hole locations.

Taps

Machine taps are designed to withstand the torque required to thread a hole and clear the chips out of the hole. Tapping is one of the most difficult machining operations to perform because of the following factors:

1) Inadequate chip clearance;
2) Inadequate supply of cutting fluid;
3) Coarse and fine threads in various materials;
4) Speed and feed of threading operations being governed by the lead of the thread;
5) Depth of thread required.

Reamers

Reamers are available in a variety of designs and sizes. A reamer is a rotary end cutting tool used to accurately size and produce a good surface finish in a hole which has been previously drilled or bored.

Boring Tools

Boring is the operation of enlarging a previously drilled, bored, or cored hole to an accurate size and location with a desired surface finish. This operation is generally performed with a single-point boring tool. When a boring bar is selected, the length and diameter should be carefully considered: as the ratio between length and diameter increases, the rigidity of the boring bar decrease. For example, a boring bar with a 1:1 length-to-diameter ratio is 64 times more rigid than one with 4:1 ratio.

New Words and Phrases

polycrystalline [ˌpɔliˈkristəlain] adj. 多晶的
reamer [ˈriːmə] n. 铰刀，扩锥
mill [mil] n. 铣床；铣刀
contour [ˈkɔntuə] n. 轮廓
tap [tæp] n. 丝锥
boring [ˈbɔːriŋ] n. 镗孔
torque [tɔːk] n. 转矩
rigidity [riˈdʒiditi] n. 刚度
carbide [ˈkɑːbaid] n. 碳化物
spot-facing adj. 刮孔口平面的
counter bore 反镗
high-speed steel 高速钢
cemented carbide 硬质合金
cubic boron nitride（CBN） 立方氮化硼
machining center 加工中心
end mill 面铣刀
shell end mill 圆筒形面铣刀（或套筒铣刀）
contour milling 轮廓铣削
boring tool 钻削刀具，镗削刀具

Unit 4

Text

Tool Radius Compensation

What is Tool Radius Compensation

Tool radius compensation is the act of accommodating for the cutting tool radius in order to produce the workpiece of the correct geometry. [1]

How Cutter Radius Compensation Works

Understanding how the CNC control interprets motions under the influence of cutter radius compensation should help you learn how to use it and avoid problems with programming. [2] In basic terms, you will tell the control (by a G code) the relationship of the milling cutter to the workpiece during the machining operation. Either the cutter will be on the left side (G41) or right side (G42) of the workpiece as it machines the contour. Once cutter radius compensation is properly instated, the control will keep the milling cutter to the left or right of all motions programmed (depending upon your choice). This can be a little difficult to visualize, especially for beginning programmers. The CNC control will automatically keep the cutter center away from the programmed path by the amount of the specified cutter radius compensation offset. [3] This is what allows you to program the work surface contour and let the control figure out the cutter's center line path.

In this example, cutter radius compensation will be instated during the movement from point one to point two (notice that point one is the cutter's center line position). Once instated, notice how the control will automatically keep the cutter to the right side of all programmed movements (Figure 4-1). How much to the right side of the work surface contour the control will keep the tool is specified in the cutter radius compensation offset (the cutter's radius). [4]

Two things are especially difficult for beginning programmers to visualize. Almost all problems a beginner will have with cutter radius compensation stem from one of these two difficulties. First is the instating movement (the movement from point one to point two). Prior to instating compensation, you must position the tool to a location larger than the cutter radius away from the first surface to be machined (point one in this example). [5] Second is the approach and retract of the tool. Notice that the first portion of the movement from point two to point three and the movement from point thirteen to fourteen are not actually part of the workpiece, yet they must be included within the work surface path.

Cutter radius compensation makes it possible to use the same work surface coordinates needed for finish milling a contour to rough the contour. [6] However, cutter radius compensation will be used only for milling cutters. Though there are some limitations, this eliminates the need for the programmer to calculate two sets of coordinates. This reason for using cutter radius compensation applies most to manual programmers.

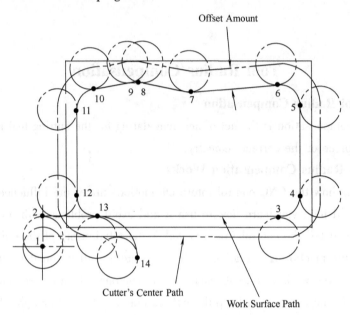

Figure 4-1　The Relationship of Cutter Center Line Path to Work Surface Path

New Words and Phrases

  finish ['finiʃ] *n.* 精加工
  rough [rʌf] *n.* 粗加工
  accommodate [əˈkɔmədeit] *vt.* 调节
  visualize [ˈviʒjuəlaiz] *vt.* 可视化,形象化;想象 *vi.* 显现
  instate [inˈsteit] *vt.* 任命,指定
  interpret [inˈtəːprit] *v.* 解释,说明;口译
  stem [stem] *vi.* 起源
  retract [riˈtrækt] *v.* 回退,缩回,缩进,所卷进(舌等);收回,取消,撤销
  figure out　合计为,计算出;解决;断定,领会到
  tool radius compensation　刀具半径补偿
  tool nose radius compensation　刀尖圆弧半径补偿
  milling cutter　铣刀

Notes

[1] Tool radius compensation is the act of accommodating for the cutting tool radius in order to produce the work piece of the correct geometry.

刀具半径补偿是一种调节刀具半径的行为，其目的是使加工出来的工件具有正确的几何形状。

accommodate 在这里是"调节"的意思，而不是我们熟悉的"提供住宿，容纳"的意思。

[2] Understanding how the CNC control interprets motions under the influence of cutter radius compensation should help you learn how to use it and avoid problems with programming.

了解 CNC 控制器在施加刀具半径补偿时如何编译运动，将十分有助于了解如何在程序中使用半径补偿并避免出错。

[3] The CNC control will automatically keep the cutter center away from the programmed path by the amount of the specified cutter radius compensation offset.

CNC 控制器会自动让刀心偏离编程轨迹，偏移的尺寸为特定的刀具半径补偿偏置。

[4] How much to the right side of the work surface contour the control will keep the tool is specified in the cutter radius compensation offset (the cutter's radius).

控制器应该让刀具处于工件表面轮廓右侧多远的地方，则由刀具半径补偿偏置（刀具半径）确定。

本句的主语是个从句——How much to the right side of the work surface contour the control will keep the tool...。

[5] Prior to instating compensation, you must position the tool to a location larger than the cutter radius away from the first surface to be machined (point one in this example).

设置指定补偿之前，必须将刀具置于一个位置，这个位置与第一个加工面间的距离应大于刀具半径（在本例的点 1 处）。

[6] Cutter radius compensation makes it possible to use the same work surface coordinates needed for finish milling a contour to rough the contour.

有了半径补偿，轮廓的粗加工和精加工就可以用相同的工件表面坐标进行编程。

Translating Skills

科技英语翻译方法与技巧——个人简历范例

样例 1：

Resume

Name：Jianping Li　　　　　　　　　　　Sex：Male

Address: No. 291 Gongnong Road Birthday: April 23, 1984
District Wancheng, Nanyang Birthplace: Nanyang
Henan Polytechnic Institute Height: 178 cm
Class 03338 Weight: 66 kg
  Health: Excellent

Postal Code: 473009
Home Address: Apt. 402, No. 12, Shanghai Road, Nanyang, 473009
ID Card Number: 411329198404232891
School Phone: (0733) 63736906
Home Phone: (0733) 63742285
Education:
September 2003 ~ July 2006, Majored in Mechanical and Electrical Technology Application at Henan Polytechnic Institute
September 1997 ~ July 2003, Nanyang No. 26 Middle School
September 1991 ~ July 1977, Nanyang Shanghai Road Primary School
(1) Electronic Courses:
Circuit Basic: 90
Digital Circuit: 88
Analogous Circuit: 92
CAD&CAM: 95
BASIC Language: 90
Technology of the Sensor and Detection: 96
PLC: 88
(2) Machinery Courses:
Technology of Machinery and Electric Control: 97
Theory of Machinery and the Machine Parts: 90
Technology of the Tolerance and Measurement: 88
Machinery Manufacturing Technology: 87
Hydrodynamic Transmission: 90
Mechanical Engineering Drawing: 90
Engineering Mechanics: 90
(3) Other Courses:
Industrial Enterprise Management: A
English for Science and Technology: A
Summer Jobs:
July 2004 ~ August 2004   Served as a Tutor
June 2005 ~ October 2005   Practitioner, Zhengzhou SHARP Electronics Corporation Ltd.

Awards:
Elected a Three Goods Student in 1998~2005.
Won the title of Excellent Leader of School Council in 2005
Special Skills:
Fluent in English Reading, Writing and Speaking
Hobbies:
Computer, Reading, Travel
References:
Will be provided upon request.

个人简历

姓名：李建平　　　　　　　　　　　　　　　　性别：男

学校地址：河南工业职业技术学院 03338 班
　　　　　南阳市宛城区工农路 291 号　　　　出生日期：1984.4.23

邮政编码：473009　　　　　　　　　　　　　出生地点：南阳

家庭地址：南阳市上海路 12 号 402 室　　　　身高：178 cm
　　　　　邮编：473009　　　　　　　　　　体重：66 kg

身份证号码：411329198404232891　　　　　　健康状况：很好

学院电话：（0377）63736906

家庭电话：（0377）63742285

学历教育：

2003.9~2006.7，河南工业职业技术学院主修机电技术应用专业

1997.9~2003.7，南阳第 26 中学

1991.9~1997.7，南阳上海路小学

(1) 电类课程：

电路基础：90 分　　数字电路：88 分　　模拟电路：92 分

CAD/CAM：95 分　　BASIC 语言：90 分　　传感与检测技术：96 分

可编程序控制：88 分

(2) 机类课程：

机电控制技术：97 分　　机械原理和机械零件：90 分　　公差测量：88 分

机制工艺：87 分　　液压传动：90 分　　机械制图：90 分　　工程力学：90 分

(3) 其他课程：

企业管理：A　　科技英语：A

暑假工作：

2004.7~2004.8　家庭教师

2005.6~2005.10　郑州夏普电子有限公司实习

奖励：

1998～2005 年被评为"三好学生"。
2005 年获得"校学生会优秀干部"称号。
特别技能：
英语读、写、说流利
业余爱好：
计算机、阅读、旅游
证明人：
如有需要将随时提供。

样例 2：

Curriculum Vitae/Resume

Name：Xiaomei Wei   Sex：Female
Address：Rm. 1605, No. 25, Dongfeng Road, Fuzhou, 235521
Telephone：(0) 75936863 (0) 7759382   E-mail：wxm@sina.com
Date of Birth：Apirl 22, 1973   Nationality：the Han
Martial Status：Single   Health：Good
Present Position：Senior Engineer in Legend Computer Ltd. in Fuzhou
Education：
1994   Bachelor of Science Wuhan University   Major：Electronic and Information
1997   Master of Science Wuhan University   Major：Computer Network
Work Experience：
9/1994～3/1996   Science teacher, the Department of Education in Wuhan University
4/1996～6/1997   Visiting Professor, University of Calagary, Alberta, Canada
7/1997～present   Senior Engineer in Legend Computer Ltd. in Fuzhou
Awards and Honors：
Work-study Scholarship in Wuhan University, 1992～1994
First-prize winner in Students Computer Contest in Wuhan University, 1993
Foreign Language：
English, passed CET-6 and fluent in spoken English

个人简历

姓名：魏晓梅   性别：女
家庭住址：福州市东风路 25 号 1605 寓所，235521
电话：(0) 75936863   (0) 77593823   电子邮箱：wxm@sina.com
生日：1973 年 4 月 22 日   民族：汉
婚姻状况：未婚   身体状况：良好
现任工作：福州市传奇计算机有限公司高级工程师

学历教育：
1994 年获得武汉大学理学学士　主修：电子与信息
1997 年获得武汉大学理学硕士　主修：计算机网络
工作经历：
1994.9～1996.3　任武汉大学教育系理科教师
1996.4～1997.6　任加拿大阿尔伯特加拉格里大学访问学者
1997.7～现在　任福州市传奇计算机有限公司高级工程师
获奖情况：
1992～1994 年获武汉大学工学奖学金
1993 年获武汉大学学生计算机竞赛一等奖
外语方面：
通过大学公共英语 6 级考试，且口语流利

Reading Material

Tool Length Offsets and Zero Presets

The different tools used for the machining operations in a program vary in length. There are two different techniques in common use in industry for correcting tool length differences. These are the use of offsets and the use of zero presets.

Tool Length Offsets

Tool length offsets are used during tool changing in a single program. Tool length offsets are used to correct the length and position of different tools so that they perform as though they had equal length and positions. The correcting offsets are called up from the tool offset registers in the controller by the tool code.

Refer to Figure 4-2, a good example would be three drills in a lathe turret. Ideally, as each tool touches the workpiece, the Z axis register would read zero. Since each tool has a different length, the Z axis offset is entered as a positive or negative adjustment. Thus, each can be programmed as though they had the same length as tool #1.

In Figure 4-2, tool #1 is the standard with which tools #2 and #3 are com-

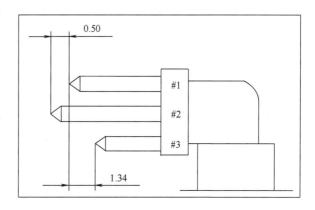

Figure 4-2　A Lathe Tool Turret Loaded with Three Tools

pared. Tool #2 is 0.50 longer than tool #1 along the Z axis. Therefore, the Z axis offset register would contain the offset of Z -0.50.

When tool code T0202 is read along with an M06 for a tool change, the machine will automatically move the tool to the compensated position—that is, put tool #2 in the same position as tool #1 in its uncompensated position. The offset will apply until is cancelled by a new tool code.

Tool #3 is shorter than tool #1. Therefore, its Z axis offset would be a positive 1.34 to bring it up to the tool #1 position.

Zero Preset

Commonly, on lathes, there is a way to set the PRZ from the departure point without actually resetting the machine axis registers. This is done by a code or word that causes the control to reset the zero point from the present location of the tool. The PRZ is established by the G50 code followed by the distance it lies from the present tool location. The control reads the G50 code followed by the coordinates of the new PRZ. The present location of the tool is taken as the temporary reference.

For example:

G50 X10 Z60

This command informs the control that the tool is presently 10 mm in X and 60 mm in Z from the PRZ. The control then knows that the PRZ for this operation is located 60 mm toward the chuck in the Z axis and on center in the X axis.

New Words and Phrases

    chuck   [tʃʌk]   *n.* 卡盘

    register   [ˈredʒistə]   *n.* 寄存器

    ideally   [aiˈdiəli]   *adv.* 理想地，在观念上地，完美地

    compensate   [ˈkɔmpənseit]   *v.* 偿还，补偿

    uncompensated   [ʌnˈkɔmpenseitid]   *adj.* 未得补偿的，没有得到赔偿的

    fashion   [ˈfæʃən]   *n.* 方式，流行，风尚，时样

    positional   [pəˈziʃənl]   *adj.* 定位的，位置的

    temporary   [ˈtempərəri]   *adj.* 暂时的，临时的，临时性

    turret   [ˈtəːrit]   *n.* 转塔刀架

    tool length offset   刀具长度偏置

    zero preset   零点预置

    tool code   刀具代码

    tool change   换刀

    program reference zero（PRZ）   程序参考零点

Part III CNC Machine Tool Structure

Unit 5

Text

MCU and CPU

The MCU is the intermediary in the total NC operation. Its main function is to take the part program and convert this information into a language that the machine tool can understand so that it can perform the functions required to produce a finished part. [1] This could include turning relays or solenoids ON or OFF and controlling the machine tool movements through electrical or hydraulic servomechanisms.

Data Decoding and Control

One of the first operations that the MCU must perform is to take the binary-coded data (BCD) from the punched tape and change it into binary digits. This information is then sent to a holding area of the MCU, which is usually called buffer storage (buffer area). The purpose of the buffer area is to allow the information or data to be transferred faster to other areas of the MCU. If there were no buffer storage, the MCU would have to wait until the tape reader decoded and sent the next set of instructions. This would cause slight pauses in the transfer of information, which in turn would result in a pause in the machine tool motion and cause tool marks in the workpiece. MCUs which do not have buffer storage must have high-speed tape readers to avoid the pauses in transferring information and the machining operation.

The data decoding and control area of the MCU processes information which controls all machine tool motions as directed by the punched tape. This area also allows the operator to make changes to the program manually, through the control panel.

MCU Development

Since the early 1950s, MCUs have developed from the bulky vacuum tube units to today's computer control units, which incorporate the latest microprocessor technology. Until the early 1970s, all MCU functions—such as tape format recognition, absolute and incremental positioning, interpolation, and code recognition—were determined by the electronic elements of the MCU. [2] This type of MCU was called hard-wired because the functions were built into the computer elements of the MCU and could not be changed.

The development of soft-wired in the mid-1970s resulted in more flexible and less costly MCUs. Simple types of computer elements, and even minicomputers, became part of the MCU. The functions that were locked in by the manufacturer with the early hard-wired systems are now included in the computer software within the MCU. This computer logic has more capabilities, is less expensive, and at the same time can be programmed for a variety of functions whenever required.

CPU

The CPU of any computer contains three sections: control, arithmetic-logic, and memory. The CPU control section is the workhorse of the computer. Some of the main functions of each part of the CPU are as follows:

1. Control Section

1) Coordinates and manages the whole computer system.

2) Obtains and decodes data from the program held in memory.

3) Sends signals to other units of NC system to perform certain operations.

2. Arithmetic-Logic

1) Makes all calculations, such as adding, subtracting, multiplying, and counting as required by the program.

2) Provides answers to logic problems (comparisons, decisions, etc.).

3. Memory

1) Provides short-term or temporary storage of data being processed.

2) Speeds the transfers of information from the main memory of the computer.

3) Has a memory register which provides a specific location to store a word and/or recall a word.

New Words and Phrases

solenoid   [ˈsəulinɔid]   n. 螺线管
buffer   [ˈbʌfə]   n. 缓冲器
pause   [pɔːz]   n. & v. 暂停
panel   [ˈpænəl]   n. 面板
incorporate [inˈkɔːpəreit] v. 合并；组合
format   [ˈfɔːmæt]   n. 形式；格式 v. 对……格式化
absolute   [ˈæbsəluːt]   adj. 绝对的
incremental [inkriˈməntl]   adj. 增加的
interpolation [inˌtəːpəuˈleiʃən] n. 插补
recall   [riˈkɔːl]   v. & n. 调用
turn on   打开

turn off 关闭
binary-coded data 二进制编码数据
binary digit 二进制数字
buffer storage 缓冲存储器
buffer area 缓冲区
tool mark 刀痕
high-speed tape reader 高速读带机
control panel 控制面板
vacuum tube 真空管

Notes

［1］Its main function is to take the part program and convert this information into a language that the machine tool can understand so that it can perform the functions required to produce a finished part.

其主要任务是接收零件程序并将这些信息转换为机床可以识别的语言，从而实现加工成品所需要的各种功能。

主句有两个并列表语 to take... 与（to）convert...，定语从句 that the machine tool can understand 修饰 a language。

［2］Until the early 1970s, all MCU functions—such as tape format recognition, absolute and incremental positioning, interpolation, and code recognition—were determined by the electronic elements of the MCU.

20 世纪 70 年代初期以前，MCU 的各项功能，如纸带格式识别、绝对定位与增量定位、插补、代码识别等，均由 MCU 的电子元器件实现。

这是一个简单句，其结构是 all MCU functions were determined by...。

Translating Skills

科技英语翻译方法与技巧——怎样写英文求职信

Your application letter is one of your most importment job-search documents. An effective letter can get you a phone call for an interview, but a poorly written application letter usually spells continued unemployment. The difference can be a matter of how you handle a few key points. The following are some tips to help you develop effective application letters.

1. Individualizing Your Letter

Give your readers some insight into you as an individual. In the example in unit 6, the writer chose to describe particular experiences and skills that could not be generalized to most other recent graduates. Draft your letter to show how your individual qualities can contribute to the organization. This is your letter, so avoid simply copying the form and style of other letters you've seen. Instead, strive to make your letter represent your individuality and your capacities.

2. Addressing a Specific Person

Preferably, the person you write to should be the individual doing the hiring for the position you're seeking. Looking for this person's name in company publications found at the university placement service. If the name is unavailable in these places, phone the organization and ask for the person's name or at least the name of the personnel manager.

3. Catching Your Reader's Attention

Your introduction should get your reader's attention, stimulate interest, and be appropriate to the job you are seeking. For example, you may want to begin with a reference to an advertisement that prompted your application. Such a reference makes your reason for contacting the company clear and indicates to them that their advertising has been effective. Or you may want to open by referring to the company's product, which you want to promote. Such a reference shows your knowledge of the company. Whatever opening strategy you use, try to begin where your reader is and lead quickly to your purpose in writing.

4. First Paragraph Tips

Make your goal clear. If you're answering an advertisement, name the position stated in the ad. and identify the source, for example: "your advertisement for a graphic artist, which appeared in the Chicago Sun Times, May 15,1988..." If you're prospecting for a job, try to identify the job title used by the organization. If a specific position title isn't available or if you wish to apply for a line of work that may come under several titles, you may decide to adapt the professional objective stated in your resume.

Additionally, in your first paragraph you should provide a preview of the rest of your letter. This tells your readers what to look for and lets him or her know immediately how your qualifications fit the requirements of the job. In the example letter in unit 6, the last sentence of the first paragraph refers to specific work experience that is detailed in the following paragraph.

5. Highlighting Your Qualifications

Organize the middle paragraphs in terms of the qualifications that best suit you for the job and the organization. That is, if your on-the-job experience is your strongest qualification, discuss it in detail and show how you can apply it to the needs of the company. Or if you were president of the marketing club and you are applying for a position in marketing or sales, elaborate on the valuable experience you gained and how you can put it to work for them. If special projects you're done apply directly to the job you are seeking, explain them in detail. Be specific. Use numbers, names of equipments you've used, or features of the project that may apply to the job you want.

One strong qualification, described so that readers can picture you actively involved on the job, can be enough. You can then refer your reader to your resume for a summary of your other qualifications. If you have two or three areas that you think are strong, you can develop additional paragraphs. Make your letter strong enough to convince readers that your distinctive background qualifies you for the job but not so long that length will turn readers off. Some employers recommend a maximum of four paragraphs.

6. Other Tips

Refer to your resume. Be sure to refer to your enclosed resume at the most appropriate point in your letter, for example, in the discussion of your qualifications or in the closing paragraph.

Conclude with a clear, courteous request to set up an interview, and suggest a procedure for doing so. The date and place for the interview should be convenient for the interviewer. However, you're welcome to suggest a range of date and place convenient to you, especially if you travel at your own expense or have a restricted schedule. Be specific about how your reader should contact you. If you ask for a phone call, give your phone number and times of the week when you can be reached.

Be professional. Make sure your letter is professional in format, organization, style, grammar, and mechanics. Maintain a courteous tone throughout to eliminate all errors. Remember that readers often "deselect" applicants because of the appearance of the letter.

Seek advice. It's always a good idea to prepare at least one draft and show to a critical reader for comments and suggestions before revising and sending the letter.

Reading Material

Machine Movements and Control

Positioning Control

The basis of numerically controlled machining is the programmed movement of the machine slides to predetermined position. The positioning is described in the following three ways.

1. Point-to-Point

Point-to-Point positioning involves programming instructions that only identify the next position required. The position may be reached by movement in one or more axes.

2. Line Motion

Line motion control is also referred to as linear interpolation. The programmed movement results from instructions that specify the next required position and also the feed rate to be used to reach that position. Linear interpolation consists of any programmed points linked by straight lines, whether the points are close together or far apart.

3. Contouring

Contouring, is the ability to control motions on two or more machine axes simultaneously to keep a constant cutter-workpiece relationship. The programmed information on the NC tape must accurately position the cutting tool from one point to the next and follow a predefined accurate path at a programmed feed rate in order to produce the form or contour required.

Loop System for Controlling Tool Movement

A loop system sends electrical signals to drive motor controllers and receives some form of electrical feedback from the motor controllers. There are two main systems in use today for controlling CNC machine movements: the open loop system and the closed loop system.

1. Open Loop System

An open loop system (Figure 5-1) utilizes stepping motors to create machine movements. These motors rotate a fixed amount, usually 1.8°, for each pulse received. Stepping motors are driven by electrical signals coming from the MCU. The motors are connected to the machine table ball-nut lead screw and spindle. Upon receiving a signal they move the table and/or spindle a fixed amount. The motor controller sends signals back indicating the motors have completed the motion. The feedback, however, is not used to check how close the actual machine movement comes to the exact movement programmed.

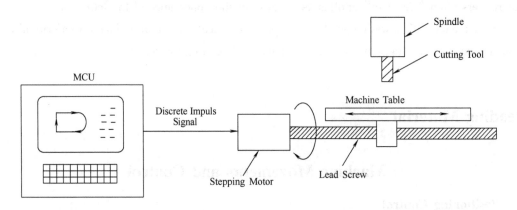

Figure 5-1 Configuration of an Open Loop System

2. Closed Loop System

Special motors called servos are used for executing machine movements in closed loop systems (Figure 5-2). Motor types include AC servos, DC servos and hydraulic servos. Hydraulic servos, being the most powerful, are used on large CNC machines. AC servos are next in strength and are found on many machining centers.

A servo does not operate like a pulse counting stepping motor. The speed of an AC or DC servo is variable and depends upon the amount of current passing through it. The speed of a hydraulic servo depends upon the amount of fluid passing through it. The strength of current coming from the MCU determines the speed at which a servo rotates.

Servos are connected to the spindle. They are also connected to the machine table through the ball-nut lead screw. A device called a revolver continuously monitors the amount by which the table and/or spindle has moved and sends this information back to the MCU. The MCU can then adjust its signal as the actual table and/or spindle position approaches the programmed position. Systems that provide feedback signals of this type are called servo systems or servo mechanisms. They can position tools with a very high degree of accuracy even when driving motors with high-horsepower ranges.

Open loop systems have recently gained renewed interest for CNC applications. Improvements in stepping motor accuracy and power have, in some cases, eliminated the need for expensive feedback system hardware and its associated circuitry. These newer systems represent substantial savings in machine and maintenance costs.

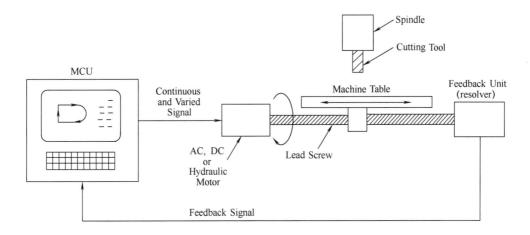

Figure 5-2　Configuration of a Closed Loop System

New Words and Phrases

slide　　［slaid］　　n. 滑板
pulse　　［pʌls］　　n. 脉冲
servo　　［ˈsəːvəu］　　n. 伺服系统
revolver　　［riˈvɔlvə］　　n. 旋转变压器
predetermine　　［ˈpriːdiˈtəːmin］　　v. 预定,预先确定
motion　　［ˈməuʃən］　　n. & v. 运动
linear　　［ˈliniə］　　adj. 线的,直线的,线性的
contouring　　［kənˈtuəriŋ］　　n. 造型
identify　　［aiˈdentifai］　　vt. 识别,鉴别;确定

instruction　[inˈstrʌkʃən]　n. 指示，指导，指令；用法说明
feedback　[ˈfiːdˌbæk]　n. 反馈
servomechanism　[ˈsəːvəuˈmekənizəm]　n. 伺服机构（系统），自动控制装置
current　[ˈkʌrənt]　n. 电流
hydraulic　[haiˈdrɔːlik]　adj. 液压的，水压的
circuitry　[ˈsəːkitri]　n. 电路，线路
substantial　[səbˈstænʃəl]　adj. 坚固的；实质的，真实的；充实的
saving　[ˈseiviŋ]　n. 节约；存款；挽救，救助
represent　[ˌriːpriˈzent]　vt. 代表；声称；表现
point-to-point control　点位控制
line motion control　直线运动控制
contouring control　轮廓控制
linear interpolation　直线插补
feed rate　进给速度
electrical signal　电信号
loop system　回路系统
open-loop system　开环系统
closed-loop system　闭环系统
stepping motor　步进电动机
ball-nut lead screw/ball screw　滚珠丝杠
feedback signal　反馈信号
AC servo　交流伺服系统
DC servo　直流伺服系统
hydraulic servo　液压伺服系统

Unit 6

Text

Types and Parts of Machining Centers

There are two main types of machining centers: the horizontal spindle and the vertical spindle machine.

1. Horizontal Spindle Type

1) The traveling-column type is equipped with one or usually two tables on which the work can be mounted. With this type of machining center, the workpiece can be machined while the operator is loading a new workpiece on the other table.

2) The fixed-column type is equipped with a pallet shuttle. The pallet is a removable table on which the workpiece has been machined. After the workpiece has been machined, the workpiece and pallet are moved to a shuttle which then rotates, bringing a new pallet and workpiece into position for machining.[1]

2. Vertical Spindle Type

The vertical spindle machining center is a saddle-type construction with sliding bedways which utilizes a sliding vertical head instead of a quill movement.

Parts of the CNC Machining Centers

The main parts of CNC machining centers are the bed, saddle, column, table, servomotors, ball screws, spindle, tool changer, and the machine control unit (MCU).

Bed—The bed is usually made of high-quality cast iron which provides for a rigid machine capable of performing heavy-duty machining and maintaining high precision.[2] Hardened and ground bedways are mounted to the bed to provide rigid support for all linear axes.

Saddle—The saddle, which is mounted on the hardened and ground bedways, provides the machining center with the X-axis linear movement.

Column—The column, which is mounted on the saddle, is designed with high torsional strength to prevent distortion and deflection during the manufacturing process. The column provides the machining center with the Y-axis linear movement.

Table—The table, which is mounted on the bed, provides the machining center with the Z-axis linear movement.

Servo system—The servo system, which consists of servo drive motors, ball screws, and position feedback encoders, provides fast and accurate movement and positioning of the XYZ axes slides. The feedback encoders mounted on the ends of the ball screws form a closed-loop system which maintains consistent high-positioning unidirectional repeatability of ±0.001 in

(0.025 4 mm).[3]

Spindle—The spindle, which is programmable in 1r/min increments, has a rotational speed from 20 to 6 000 r/min. The spindle can be of a fixed position (horizontal) type, or can be a tilting/contouring spindle which provides for an additional (A) axis.[4]

Tool changers—There are basically two types of tool changers, the vertical tool changer and the horizontal tool changer. The tool changer is capable of storing a number of preset tools which can be automatically called for use by the part program. Tool changers are usually bidirectional, which allows for the shortest travel distance to randomly access a tool. The actual tool change time is usually only 3 to 5 s.

MCU—The MCU allows the operator to perform a variety of operations such as programming, machining, diagnostics, tool and machine monitoring, etc. MCUs vary according to manufacturers' specifications; new MCUs are becoming more sophisticated, making machine tools more reliable and the entire machining operations less dependent on human skills.

New Words and Phrases

equip	[iˈkwip]	v.	装备，配备
mount	[maunt]	n.	装备 v. 安装；设置；安放
pallet	[ˈpælit]	n.	平板架
shuttle	[ˈʃʌtl]	n.	滑闸，滑台 v. 穿梭往返
saddle	[ˈsædl]	n.	鞍，鞍状物
bedway	[ˈbedwei]	n.	床身导轨
quill	[kwil]	n.	衬套；主轴
harden	[ˈhɑːdn]	v.	使变硬；淬火
strength	[streŋθ]	n.	力；强度
distortion	[disˈtɔːʃən]	n.	扭曲，变形；失真
deflection	[diˈflekʃən]	n.	偏斜，偏转；偏差
encoder	[inˈkəudə]	n.	编码器
increment	[ˈinkrimənt]	n.	增加；增量
bidirectional	[ˌbaidiˈrekʃənəl]	adj.	双向的

traveling-column　移动立柱
fixed-column　固定立柱
ball screw　滚珠丝杠
tool changer　换刀机构
machine control unit (MCU)　机床控制单元
cast iron　铸铁
torsional strength　扭转强度

Notes

［1］After the workpiece has been machined, the workpiece and pallet are moved to a shuttle which then rotates, bringing a new pallet and workpiece into position for machining.

完成加工后，工件与平板架一起移到滑台，然后滑台转动，把新的平板架与工件输送到加工位置。

定语从句 which then rotates 修饰 a shuttle。

［2］The bed is usually made of high-quality cast iron which provides for a rigid machine capable of performing heavy-duty machining and maintaining high precision.

床身通常由优质铸铁制成，使机床具有刚性加工能力，可实现重型加工，并能保持良好的加工精度。

which 引导的定语从句修饰 high-quality cast iron。

［3］The feedback encoders mounted on the ends of the ball screws form a closed-loop system which maintains consistent high-positioning unidirectional repeatability of ±0.001 in (0.025 4 mm).

反馈编码器安装在滚珠丝杠端部，构成闭环系统，单向重复定位精度可达±0.001英寸（0.025 4 毫米）。

主句结构为 The feedback encoders... form a closed-loop system...。过去分词短语 mounted on the ends of the ball screws 作后置定语修饰 encoders；which 引导的定语从句修饰 a closed-loop system。

［4］The spindle can be of a fixed position (horizontal) type, or can be a tilting/contouring spindle which provides for an additional (A) axis.

主轴可以是固定式（卧式），或是倾斜/轮廓主轴，以提供一个附加轴（A 轴）。

主句谓语采用 can be of 的形式，be of 作谓语常表示事物的属性。

Translating Skills

科技英语翻译方法与技巧——英文求职信范例

样例1：

Dear Ms. Rennick,

　　Dr. Saul Wilder, a consultant to your firm and my Organizational Management professor, has informed me that Aerosol Monitoring and Analysis is looking for someone with excellent communication skills, organization experience, and leadership background to train for a management position. I believe that my enclosed resume will demonstrate that I have the characteristics and experience you seek. In addition, I'd like to mention how my work experience last summer makes me a particularly strong candidate for the position.

As a promoter for Kentech Training on the 1997 Paris Air Show, I discussed Kentech's products with marketers and sales personnel from around the world. I also researched and wrote reports on new product development and compiled information on aircraft industry trends. The knowledge of the aircraft industry I gained from this position would help me analyze how Aerosol products can meet the needs of regular and prospective clients, and the valuable experience I gained in promotion, sales, and marketing would help me use that information effectively.

I would welcome the opportunity to discuss these and other qualifications with you. If you are interested, please contact me at (317) 555-0118 any morning before 11:00 a.m., or feel free to leave a message. I look forward to meeting with you to discuss the ways my skills may best serve Aerosol Monitoring and Analysis.

Sincerely yours,
(Signature)
Name (Typed)

Enclosure: resume

尊敬的 Rennick 女士：

贵公司的顾问、我的组织管理教授 Saul Wilder 博士通知我，气雾剂监管分析公司正在招聘一位交际能力好、拥有组织经验及领导背景的人，经培训可以就任管理岗位。相信我内附的履历会证明我拥有您期待的特征与经历。另外，去年夏天的工作经历会证明我正是这一职位的强有力的候选人。

作为1997年巴黎航空展 Kentech 培训的发起人，我和来自世界各地的市场商人和销售人员讨论过 Kentech 的产品。我还曾研究并就新产品的开发以及汇编飞机制造工业趋势方面的信息写出了报告。从这一职位所获得飞机制造工业的知识将帮助我分析如何能使气雾剂产品满足新老客户的需求，在产品的宣传、销售以及推向市场的过程中所获得的宝贵经验将帮助我有效地使用这些信息。

希望能够有机会与您讨论上述的以及其他的资历。如果您感兴趣，请在上午11:00以前拨打（317）555-0118 与我联系，或者留下口信。盼望着与您会面并讨论如何使我的技能能最好地服务于气雾剂监管分析公司。

谨启

申请人签名

随信附上：履历

样例2：

Dear _____,

The need for a biology teacher in the Heavilon Community Schools was indicated in the Purdue University Educational listing of March 7, 2009. If this vacancy still exists, please consider me as an applicant for the position and send me a teacher application.

On May 13, 2009, I will graduate from Purdue University with a B. S. degree. I will receive an Indiana Secondary Standard Certificate with a major in biology and a minor in botany. In addition to this formal education, the past two summer were spent at the Gull Lake Biological Station where I worked as a laboratory assistant. The preceding two summers were spent in field work with the Indiana Department of Conservation. These experiences have been valuable additions to my educational background. During the fall semester of 2003, I did my student teaching in biology at Central High School in Lafayette, Indiana.

My resume is enclosed for your information, and my credentials are available at the Educational Placement Office, Purdue University, West Lafayette, Indiana. A personal interview can also be arranged at your request.

Sincerely,
(Signature)
Name (Typed)

Enclosure: Resume

尊敬的_____：

我在2009年3月7日的普渡大学教育名单上看到海威龙社区学校招聘生物老师的信息。如果这个职位仍然空缺，请将我列为此职位的申请人并请寄给我一份教师申请表。

我将于2009年5月13日从普渡大学毕业并获得理学学士学位，我将收到一份主修生物学、兼修植物学的印第安纳州二级水平证书。除此正规教育之外，我在过去的两个夏天都在高尔湖生物学基地当实验员。再往前的两个夏天，我和印第安纳州自然保护部门一起做调查工作。这些经历都是我教育背景珍贵的补充。2003年秋季学期，我在印第安纳州拉法耶特市中心学校进行生物学教学实习。

随信寄去我简历，在印第安纳州西拉法耶市普渡大学的教育就业指导处可以获得我的资格审查。我将应您的要求，参加面试。

谨启

申请人（签名）

随信附上：个人简历

Reading Material

The Axis System

What Is a Machine Axis?

Axes are the plural of axis. They both mean a reference for motion and position. An axis is a central line to which all CNC movement is compared or referenced. The axis system is a worldwide standard for machine movement.

Motion and Direction

Axes are used to identify motion. All CNC machines require some method of identifying which motion is needed in the program. For example, on a vertical milling machine, if the programmer requires the left and right movement of the tools, the X-axis would be used; similarly, the in and out motion of the table would require the Y-axis.

Direction is determined by sign value. Once the axis is identified, the sign of the axis, plus or minus, determines the direction of its movement: left or right, in or out, up or down, clockwise or counterclockwise.

Refer to the milling machine in Figure 6-1, the X-axis of this milling machine table is the left and right movement. If the programmer needs a tool movement to the right, he or she would call out a "plus X movement". A movement to the left would be "minus X". Movement of the tool inward (away from the operator) is plus Y, and tool-up would be plus Z.

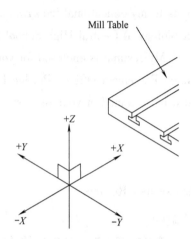

Figure 6-1 The Three Axes of a Vertical Milling Machine

The Axis Framework

Note that the three axes on the mill (Figure 6-1) are at 90° to each other. This is called an orthogonal axis frame. Orthogonal means "at 90°". The standard axes on most CNC machines are orthogonal.

Right-hand Rule

You can identify the axis framework on most CNC machines by the right-hand rule illustrated in Figure 6-2. If the thumb of your right hand points along the positive X-axis, the first finger will point out the positive Y-axis and the second finger will identify the positive Z-axis.

There are two kinds of single machine axis movement. Machine axis motion is either in a straight line (linear movement) or in a circle (rotary movement). Each depends upon an axis for reference.

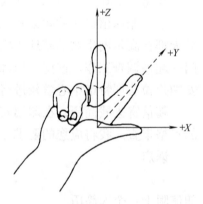

Figure 6-2 The Right-hand Rule Shows the Relationship of the Primary Linear Axes of CNC Machines

An axis is a straight line used for reference for motion and distance. Rotary motion occurs around this central (symmetrical) axis. This is similar to the motion of a wheel around an axle. Linear motion is parallel to the reference axis line.

New Words and Phrases

axis　　['æksis]　　*n.* 轴
plural　　['pluərəl]　　*adj.* 复数的
reference　　['refrəns]　　*n.* 参考
orthogonal　　[ɔː'θɔɡənl]　　*adj.* 直角的，正交的
framework　　['freimwəːk]　　*n.* 构件；框架，结构
rotary　　['rəutəri]　　*adj.* 旋转的
symmetrical　　[si'metrikəl]　　*adj.* 对称的，均匀的
axle　　['æksl]　　*n.* 轮轴，车轴
vertical milling machine　　立式铣床
axis framework　　坐标系
orthogonal axis frame　　直角坐标系
right-hand rule　　右手法则
linear movement/motion　　直线运动
rotary movement/motion　　旋转运动

Part IV   CNC Machine Tool System

Unit 7

Text

FANUC System Operation Unit—CRT/MDI Panel

Figure 7-1 shows standard keyboard sheet on MDI Panel.

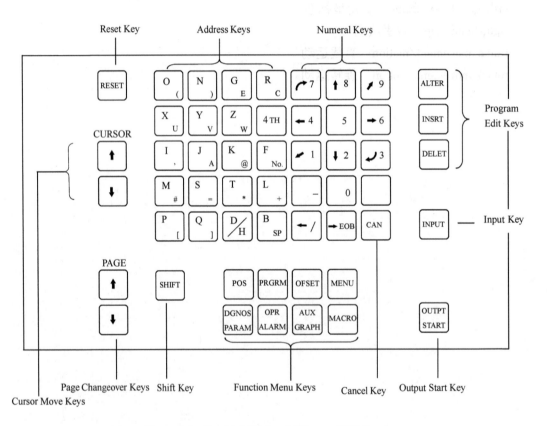

Figure 7-1   Standard Keyboard Sheet on MDI Panel

The following are MDI keyboard functions:

1. Power ON/OFF Key

This button turns the CNC power on and off.

2. Reset Key

Press this key to reset the CNC, to cancel an alarm, etc.

3. Address and Numerical Keys

Press these keys to input letters, numbers, and other characters.

4. Function Menu Keys

Position Key: Display machine tool current position on the CRT screen.

Program Key: In edit mode, edit and display the program which is in the memory. In MDI mode, input and display the MDI data.

Offset Key: Tool offset data write mode selection, mode is tool offset setting.

Menu Key: Display the operation menu.

Diagnose / Parameter Key: Set and display self-diagnoses list and parameter list.

Operation / Alarm Key: Display alarm number.

Auxiliary / Graphic Key: Make graphic simulation.

Macro Key: Set macro function.

5. Shift Key

Some address keys have two characters on top. Press the shift key switches the characters. When the shift key is pressed, "-" is displayed in the key input buffer. This indicates that the character on the lower right side of the address key can be input by pressing the address key.[1]

6. Cursor Move Keys

There are two kinds of cursor move key described below.

→: This key is used to shift the cursor a short distance in the forward direction.

←: This key is used to shift the cursor a short distance in the reverse direction.

7. Page Changeover Keys

Two kinds of page changeover keys are described below.

↓ : This key is used to changeover the page on the CRT screen in the forward direction.

↑ : This key is used to changeover the page on the CRT screen in the reverse direction.

8. Cancel Key

Press this key to delete the last character or symbol input to the key input buffer. The contents of the key input buffer are displayed on the CRT screen. When the address key or the numerical key is pressed next, the position where the alphabet or numerical is inserted next is indicated by "-". When cancel (CAN) key is pressed, the character immediately before "-" is canceled.

9. Program Edit Keys

Alter Key: Press this key to modify the last character or symbol input to the key input buffer.

Insert Key: Insert character or symbol before the current cursor.

Delete Key：Delete specific character symbol or program.

10. Input Key

When an address or a numerical key is pressed, the alphabet or the numeral is input once to the key input buffer, and is displayed on the CRT screen. To set the data input the key input buffer in the offset register, etc, press the INPUT key.

11. Output Start Key

After Press this key, CNC start to output program or parameter from the memory to peripheral equipment.

New Words and Phrases

  cursor   [ˈkəːsə]   n. 光标
  cathode ray tube (CRT)   阴极射线管
  manual data input (MDI)   手动数据输入
  peripheral equipment   外部设备

Notes

[1] This indicates that the character on the lower right side of the address key can be input by pressing the address key.

这表明通过按下地址键，按键右下角的字符能被输入。

that 引导宾语从句，从句中主语是 the character，谓语是 can be input。

Translating Skills

科技英语翻译方法与技巧——it 的用法

1. it 作代词

1) it 作无人称代词。

it 作无人称代词时可以表示自然现象、天气、时间、距离。此时 it 是形式上的主语，没有词汇意义，翻译时可省略。

2) it 作人称代词。

it 作人称代词时，用来代替上文中提到的事或物，翻译时可译为它所代替的事或物。例如：

The zener diode maintains the voltage across its terminals by varying the current that flows through it.

稳压二极管通过改变流过它的电流来维持其端电压。（这里的 it 是指稳压二极管）

2. it 作形式主语

it 作形式主语时，可代替不定式短语或主语从句。这时 it 称为形式主语。

1）it 代替不定式的句型有：

It is（was）+形容词+不定式 或 It + 谓语动词 + 不定式

例如：

It requires power to drive machines.

开动机器需要能源。

2）it 代替主语从句的句型有：

It is + 形容词 + 主语从句

It is certain that... 可以肯定的是…… It is desirable that... 理想的是……

It + 谓语动词的被动态 + 主语从句

It is reported that... 据报道…… It is generally recognized that... 大家公认为……

It is supposed that... 假设…… It is assumed that... 假定……

例如：

It has been found that a force is needed to change the motion of a body.

人们发现，要改变一个物体的运动（状态）需要加外力。

It is + 名词 + 主语从句

It is a pity that... 遗憾的是…… It is common knowledge that... 人所共知是……

It + 不及物动词 + 主语从句

It seems that... 好像是…… It turns out that... 显然，……

It now appears that... 现在看来……

3. it 作形式宾语

当不定式短语或从句在句中作宾语，而这种宾语又带有补足语时，通常要把 it 放在宾语补足语的前面，使语句简洁明了。

1）it 代替动词不定式短语。例如：

When we want to measure very small currents we find it convenient to use milliampere and microampere.

当要测量很小的电流时，我们发现用毫安和微安是比较方便的。

2）it 代替宾语从句。例如：

The effects we have just discussed make it apparent that there is a means of converting mechanical energy into electrical energy.

由刚才的讨论结果可以很明显地看出，存在一种将机械能转化为电能的方法。

4. it 用于强调句型

强调句型是简单句，可以用来强调句中的主语、宾语和状语，但不能强调谓语和定语。强调语句的句型为：It +（was）+被强调的成分+that（who）...。在这种句型中，it 和 that 都没有意义，翻译时可在强调成分前加上"正是"、"就是"等。例如：

It is in the form of alloys that metals are often used in industry.

在工业中就经常以合金的形式使用金属。

Reading Material

FANUC System Operator's Panel

Figure 7-2 shows the front view of FANUC system operator's panel, and the meanings of each key are shown in Table 7-1.

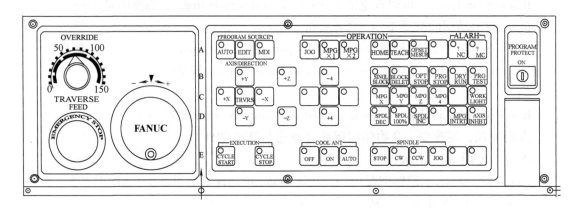

Figure 7-2 The Front View of Operator's Panel

Table 7-1 Meanings of the Keys

Keys	Meanings
AUTO	AUTO mode selection: Mode is automatic.
MDI	MDI mode selection: Mode is MDL.
EDIT	EDIT mode selection: Mode is program edition.
JOG	JOG feed mode selection: Mode is jog feed.
IN JOG	STEP feed mode selection: Mode is step feed.
MPG	MANDLE fed mode selection: Mode is manual handle feed.
HOME	ZERO return mode selection: Mode is zero return.
TEACH	TEACH in JOG (TEACH in HANDLE) MODE selection: Mode is teach in jog (teach in handle).
OFSET MESUR	TOOL OFFSET DATA write Mode selection: Mode is tool offset setting.
? NC	CNC alarm: LED is turned on when CNC alarm. Key pad does not have any meanings.
? MC	MACHINE alarm: LED is turned on when Machine alarm. Key pad does not have any meanings.

Unit 7

(续)

Keys	Meanings
SINGL BLOCK	Single block: Single block program execution for test operation.
PRG STOP	PROGRAM STOP (output only): When automatic is stopped by M00 command in the program. LED on the button is turned on.
OPT STOP	OPTIONAL STOP: When this single is turned on, auto running operation is stopped after executing M01 block.
DRY RUN	DRY RUN: When this signal is turned on, axes feed will be JOG feed speed, not command speed on the program. To check the moving of tool without workpiece.
PRG TEST	MACHINE LOCK: When this signal is turned on and auto funning operation, axes are not moved but position screen on CRT is only executed. To check the program.
MPG X	HANDLE feed X axis selection: When manual handle feed mode is selected. To check the program. (Same for Y-axis, Z-axis, C-axis and 4-axis).
WORK LIGHT	WORK LIGHT: WORK LIGHT ON\OFF control.
MPG INTRT	HANDLE INTERRUPTION: Selecting this button in automatic operation, the moving value of manual handle feed is added to the moving value of program.
AXIS INHBT	AXES INHIBIT: The specific axis or all axes are stopped to move.
LOW ×1	JOG (STEP) FEED OVERRIDE UPPER: Set override to the jig feed (or step feed) five steps display.
MEDL ×10	HANDLE FEED MULTIPLY LOWER: Manual handle multiply ×1, ×10, ×100, ×1000.
+X	MANUAL FEED DIRECTION: Selecting this button in jog feed (or step feed), selected axis moved to the selected direction by jog feed (or stop feed). (Same for −X, +Y, −Y, and +Z)
TRVRS	TRAVERSE: Executing jog in on this button, the jog feed is executed in the rapid traverse.
CYCLE START	CYCLE START: Automatic operation is started.
CYCLE STOP	CYCLE STOP: Automatic operation is stopped.
CLNT ON	COOLANT ON: Coolant is started.
CLNT OFF	COOLANT OFF: Coolant is stopped.
CLNT AUTO	COOLANT AUTO: Coolant ON\OFF control in AUTO operation.
SPDL 100%	SPINDLE OVERRIDE 100%: The rotation of spindle motor is 100% override.
SPDL INC	SPINDLE SPEED ACCELERATION: The rotation of spindle motor is accelerated.
SPDL DEC	SPINDLE SPEED DECELERATION: The rotation of spindle motor is decelerated.

（续）

Keys	Meanings
SPDL CW	SPINDLE DIRECTION CW: The rotating direction of spindle motor is clockwise.
SPDL CCW	SPINDLE DIRECTION CWW: The rotating direction of spindle motor is counterclockwise.
SPDL STOP	SPINDLE STOP: The rotation of spindle motor is stopped.
SPDL JOG	SPIDLE MANUAL FEED: SPINDLE FEED is manually.

Unit 8

Text

Servo Controls

Serve controls can be group of electrical, hydraulic, or pneumatic devices which are used to control the position of machine tool slides. The most common servo control systems in use are the open-loop and the closed-loop systems.

In the open-loop systems, the tape is fed into a tape reader which decodes the information punched on the tape and stores it briefly until the machine is ready to use it. The tape reader then converts the information into electrical pulses or signals. These signals are sent to the control unit, which energizes the servo control units. The servo control units direct the servomotors to perform certain functions according to the information supplied by the tape. The amount each servomotor will move depends upon the number of electrical pulse it receives from the servo control unit. Precision lead screws, usually having 10 threads per inch (tpi), are used on NC machines. If the servomotor connected to the lead screw receives 1 000 electrical pulses, the machine slide will move 1 in (25.4 mm). Therefore, one pulse will cause the machine slide to move 0.001 in. (0.025 4 mm). The open-loop system is fairly simple; however, since there is no means of checking whether the servomotor has preformed its function correctly, it is nor generally used where accuracy greater than 0.001 in. (0.025 4 mm) is required.

The open-loop system may be compared to a gun crew that has made all the calculations necessary to hit a distant target but does not have an observer to confirm the accuracy of the shot.

The closed-loop system can be compared to the same gun crew that now has an observer to confirm the accuracy of the shot. The observer relays the information regarding the accuracy of the accuracy of the shot to the gun crew, which then makes the necessary adjustments to hit target.

The closed-loop system is similar to the open-loop system with the exception that a feedback unit is introduced into the electrical circuit. This feedback unit, often called a transducer, compares the amount the machine table has been moved by the servomotor with the signal sent by the control unit. The control unit instructs the servomotor to make whatever with the signal sent by the control unit. The control unit instructs the servomotor to make whatever adjustments necessary until both the signal from the control unit and the one from the servo unit are equal. In the closed-loop system, 10 000 electrical pulses are required to move the machine slide 1 in (25.4 mm). Therefore, on this type of system, one pulse will cause a 0.000 1 in (0.002 54 mm) movement of the machine slide. Closed-loop NC systems are very accurate because the command signal is recorded, and there is an automatic compensation for error. If the machine slide is forced out of position due to cutting forces, the feedback unit indicates this movement and the machine control unit (MCU) automatically makes the necessary adjustments to bring the machine slide back to position.

New Words and Phrases

electrical	[iˈlektrikəl]	*adj.*	电的；用电的
hydraulic	[haiˈdrɔːlik]	*adj.*	水的；液压的
pneumatic	[njuːˈmætik]	*adj.*	风力的；气压的
decode	[diːˈkəud]	*v.*	解码，译码
convert	[kənˈvəːt]	*v.*	变换，转换
pulse	[pʌls]	*n.*	脉冲
signal	[ˈsignəl]	*n.*	信号 *v.* 发信号
send	[send]	*v.*	发送
energize	[ˈenədʒaiz]	*v.*	提供能量
direct	[diˈrekt]	*v. & n.*	指挥；命令
servomotor	[ˈsəːvəuˈməutə]	*n.*	伺服电动机
receive	[riˈsiːv]	*v.*	接收
precision	[priˈsiʒən]	*n.*	精度
thread	[θred]	*n.*	（螺纹的）头数
relay	[riˈlei]	*v.* 传送 *n.*	继电器
transducer	[trænsˈdjuːsə]	*n.*	传感器；变频器
instruct	[inˈstrʌkt]	*n.* 指令 *v.*	命令
command	[kəˈmɑːnd]	*n.* 命令 *v.*	命令
compensation	[ˌkɔmpənˈseiʃən]	*n.*	补偿

lead screw 丝杠
servo control 伺服控制
servo control unit 伺服控制单元
electrical circuit 电路
electrical device 电子设备
pneumatic device 气动设备
open-loop system 开环系统
closed-loop system 闭环系统
feedback unit 反馈单元
cutting force 切削力

Translating Skills

科技英语翻译方法与技巧——英汉习语的文化差异及翻译方法

习语是某一语言在使用过程中形成的独特、固定的表达方式。本文所要讨论的习语是广义的，包括成语、谚语、歇后语、典故等。英汉两种语言历史悠久，包含着大量的习语，它们或含蓄、幽默，或严肃、典雅，不仅言简意赅，而且形象生动、妙趣横生，给人一种美的享受。由于地理、历史、宗教信仰、生活习俗等方面的差异，英汉习语承载着不同民族的文化特色和文化信息。它们与文化传统紧密相连，不可分割。习语中的文化因素往往是翻译中的难点。本文试图借助语用学的理论，对英汉习语的翻译作一些探索。

朱光潜先生在《谈翻译》一文中说："外国文学最难了解和翻译的第一是联想的意义……"，"它带有特殊的情感氛围，甚深广而微妙，在字典中无从找出，对文学却极要紧。如果我们不熟悉一国的人情风俗和文化历史背景，对于文字的这种意义就茫然，尤其是在翻译时这种字义最不易应付。"英国文化人类学家爱德华·泰勒在《原始文化》（1871）一书中，首次把文化作为一个概念提了出来，并表述为："文化很复杂，它包括知识、信仰、艺术、道德、法律、风俗以及社会上习得的能力与习惯。"可见文化的覆盖面很广，它是一个复杂的系统。语言作为文化的一个组成部分，反映一个民族丰富多彩的文化现象。我们经过归纳总结认为，英汉习语所反映的文化差异主要表现在以下几个方面：

1. 生存环境的差异

习语的产生与人们的劳动和生活密切相关。英国是一个岛国，历史上航海业曾一度领先世界；而汉民族在亚洲大陆生活繁衍，人们的生活离不开土地。比喻"花钱浪费，大手大脚"，英语是 spend money like water，而汉语是"挥金如土"。英语中有许多关于船和水的习语，在汉语中没有完全相同的对应习语，如 to rest on one's oars（暂时歇一歇），to keep one's head above water（奋力图存），all at sea（不知所措）等。

在汉语的文化氛围中，"东风"即是"春天的风"，夏天常与酷暑炎热联系在一起，"赤日炎炎似火烧"、"骄阳似火"是常被用来描述夏天的词语。而英国地处西半球、北温带，是海洋性气候，因此报告春天消息的是西风，如英国著名诗人雪莱的《西风颂》正是对春的讴歌。英国的夏季正是温馨宜人的季节，常与"可爱"、"温和"、"美好"相连。莎士比亚在他的一首十四行诗中把爱人比作夏天——Shall I compare thee to a summer's day? / Thou art more lovely and more temperate。

2. 习俗差异

英汉习俗差异是多方面的，最典型的莫过于在对狗这种动物的态度上。狗在汉语中是一种卑微的动物。汉语中与狗有关的习语大都含有贬义，如"狐朋狗党"、"狗急跳墙"、"狼心狗肺"、"狗腿子"等，尽管近些年来养宠物狗的人数大大增加，狗的"地位"似乎有所改变,但狗的贬义形象却深深地留在汉语言文化中。而在西方英语国家,狗被认为是人类最忠诚的朋友。英语中有关狗的习语除了一部分因受其他语言的影响而含有贬义外,大部分都没有贬义。在英语习语中,常以狗的形象来比喻人的行为。如 You are a lucky dog(你是一个幸运儿),Every dog has his day(凡人皆有得意日),Old dog will not learn new tricks(老人学不了新东西)等。形容人"病得厉害"用 sick as a dog,"累极了"是 dog-tired。与此相反,中国人十分喜爱猫,用"馋猫"比喻人贪嘴,常有亲昵的成分,而在西方文化中,猫被用来比喻"包藏祸心的女人"。

3. 宗教信仰方面

与宗教信仰有关的习语也大量地出现在英汉语言中。佛教传入中国已有近两千年的历史,古人相信有"佛祖"在左右着人世间的一切,因此流传下来的与此有关的习语很多,如"借花献佛"、"闲时不烧香,临时抱佛脚"等。在西方许多国家,特别是在英美,很多人信奉基督教,相关的习语如 God helps those who help themselves(上帝帮助自助的人),也有 Go to hell(下地狱去)这样的诅咒。

4. 历史典故

英汉两种语言中还有大量由历史典故形成的习语,这些习语结构简单、意义深远,往往是不能单从字面意义去理解和翻译的。如汉语中的"东施效颦"、"名落孙山"、"叶公好龙"等。英语典故习语多来自《圣经》和希腊罗马神话,如 Achilles' heel(唯一致命弱点)、meet one's Waterloo(一败涂地)、a Penelope's web(永远完不成的工作)、a Pandora's box(潘多拉之盒,表示灾难、麻烦、祸害的根源)等。

Reading Material

FANUC-BESK NC System

The FANUC-BESK system 3M-Model A is a 2 or 3-axis contouring control CNC for drilling and milling machine, In particular, the system 3M-Model A is designed to offer maximum economy. Therefore, it is most suitable for numerical control of conventional milling machines.

By the use of a high-speed microprocessor and many custom LSIs in the control circuits, the number of circuit elements has been drastically reduced, and the number of digital logic printed circuit boards has been reduced to only one. The result is outstandingly high reliability. At the same time, since the control section including a power supply is very small, it can be easily incorporated into a power-magnetic control cabinet at machine side.

The FANUC-BESK system 3M-Model A is of pane-mount structure so that it can easily realize the integration of electric and mechanical systems. Furthermore, a built-in programmable controller is available so that power-magnetic control circuitry is simplified, realizing a more compact and integrated NC machine system.

Also available are the FANUC-BESK DC Servo Motors which have been widely used throughout the world, providing high-speed, powerful and stable machining. Furthermore, the FANUC-BESK DC Spinal Motor is capable of electric spindle orientation control. It is therefore more useful than ever.

AXIS Control by Manual Pulse Generators and Playback

Using three manual pulse generators, a machine can be manually operated by simultaneous 2-axis control, like a conventional milling machine. Playback is possible the result of a manual trial cutting can be memorized in the NC for automatic operation.

Multiplication of the manual pulse generator is possible up to 100. So the operator can manually move a machine in the same manner as with the handle of a conventional milling machine.

High Performance, High Reliability

Extensive use of high-speed microprocessors and many custom LSIs in the control circuit has drastically reduced the number of circuit elements, and only one digital logic board is used for the control section

Highly efficient power supply has been employed to reduce the generation. The keyboard switches for the NC operator's panel incorporate special rubber covers for dust. Even if a drift in the servo-loop occurs, it is automatically compensated to maintain accurate positioning. Furthermore, very careful selection of components and very extensive performance tests before shipment ensure long-lasting trouble-free operation.

Easier Maintenance

Maintenance of the FANUC-BESK System 3m- Model A is mush easier, the reasons are listed below:

1) The microprocessor always monitors the internal state of the NC and the state can be classified and displayed. When a failure should happen, an alarm lamp lights and the NC stops its operation, and the detailed cause is classified and displayed.

2) All ON / OFF signals going into and out of the NC can be shown on the display.

3) Any ON/OFF signals going out of the NC can be issued manually through the manual date input in bit-bit manner.

4) Various preset parameters such as acceleration/ deceleration time constant and rapid traverse speed can be shown on the display.

Part V CNC Programming

Unit 9

Text

Programming Concepts

Before you can fully understand CNC, you must first understand how a manufacturing company processes a job that will be produced on a CNC machine. The following is an example of how a company may breakdown the CNC process.

Flow of CNC Processing

1. Obtain or develop the part drawing.
2. Decide what machine will produce the part.
3. Decide on the machining sequence.
4. Choose the tooling required.
5. Do the required math calculations for the program coordinates.
6. Calculate the speeds and feeds required for the tooling and part material.
7. Write the NC program.
8. Prepare setup sheets and tool lists.
9. Send the program to the machine.
10. Verify the program.
11. Run the program if no changes are required.

Preparing a Program

A program is a sequential list of machining instructions for the CNC machine to execute. These instructions are CNC codes that contain all the information required to machine a part, as specified by the programmer.[1]

CNC codes consist of blocks (also called lines), each of which contains an individual command for a movement or specific action. As with conventional machines, one movement is made before the next one. This is why CNC codes are listed sequentially in numbered blocks.

The following is a sample CNC milling program. Note how each block is programmed and usually contains only one specific command. Also note that the blocks are numbered in increments of 5 (this is the software default on startup). Each block contains specific information for the machine to execute in sequence.

 Workpiece Size: X4, Y3, Z1
 Tool: Tool #3, 3/8″ Slot Drill
 Tool Start Position: X0, Y0, Z1.0

%	(Program Start Flag)
: 1002	(Program #1002)
N5 G90 G20 G40 G17	(Block #5, Absolute in Inches)
N10 M06 T3	(Tool Change to Tool #3)
N15 M03 S1250	(Spindle on CW at 1250 RPM)
N20 G00 X1.0 Y1.0	(Rapid over to X1.0, Y1.0)
N25 Z0.1	(Rapid down to Z0.1)
N30 G0 Z-0.125 F5	(Feed down to Z-0.125 at 5 ipm)
N35 X3.0 Y2.0 F10.0	(Feed diagonally to X3.0, Y2.0 at 10 ipm)
N40 G00 Z1.0	(Rapid up to Z1.0)
N45 X0 Y0	(Rapid over to X0, Y0)
N50 M05	(Spindle Off)
N55 M30	(Program End)

CNC Codes

There are two major types of CNC codes, or letter addresses, in any program. The major CNC codes are called G-codes and M-codes.

G-codes are preparatory functions, which involve actual tool moves (for example, control of the machine). These include rapid moves, feed moves, radial feed moves, dwells, roughing, and profiling cycles.

M-codes are miscellaneous functions, which include actions necessary for machining but not those that are actual tool movements (for example, auxiliary functions). These include actions such as spindle on and off, tool changes, coolant on and off, program stops, and related functions.

Other letter addresses are variables used in the G-and M-codes to make words. Most G-codes contain a variable, defined by the programmer, for each specific function. Each designation used in CNC programming is called a letter address.

The letters used for programming are as follows:

 N Block number: specifies the start of a block
 G Preparatory function, as previously explained
 X X-axis coordinate
 Y Y-axis coordinate
 Z Z-axis coordinate
 I X-axis location of arc center

J   Y-axis location of arc center
K   Z-axis location of arc center
S   Sets the spindle speed
F   Assigns a feedrate
T   Specifies tool to be used
M   Miscellaneous function, as previously explained
U   Incremental coordinate for X-axis
V   Incremental coordinate for Y-axis
W   Incremental coordinate for Z-axis

New Words and Phrases

breakdown　　['breik‚daun]　　n. 崩溃；衰弱；细目分类
sequence　　['siːkwəns]　　n. 次序，顺序，序列
sheet　　[ʃiːt]　　n. （一）片，（一）张；清单
verify　　['verifai]　　vt. 检验，校验，查证，核实
sequential　　[si'kwinʃəl]　　adj. 连续的；有序的；结果的
execute　　['eksikjuːt]　　vt. 执行；实行
contain　　[kən'tein]　　vt. 包含，容纳，容忍
individual　　[‚indi'vidjuəl]　　n. 个人，个体　adj. 个别的，单独的，个人的
increment　　['inkrimənt]　　n. 增加，增量
spindle　　['spindl]　　n. 轴，杆，心轴
profile　　['prəufail]　　n. 剖面，侧面；外形，轮廓
consist of　　包括
miscellaneous / auxiliary function　　辅助功能

Notes

[1] These instructions are CNC codes that contain all the information required to machine a part, as specified by the programmer.

这些加工指令是一些 CNC 代码，包含了加工零件的全部信息，由程序员编写。

that 引语从句，修饰 CNC codes, required to machine a part 过去分词短语修饰 the information。

Translating Skills

科技英语翻译方法与技巧——如何阅读电子产品的英文说明书

在社会生活中，人们经常见到一些电子产品的说明书。如果从基础英语的语法角度来理解，读者往往感到词不达意，特别是一些进口电子产品的说明书，理解的难度会更大。因此，人们应该了解电子产品英文说明书的特点，从而更好地阅读这些内容。

一、电子产品说明书的特点

电子产品说明书主要用来指导用户使用、维护和保管产品，以免因使用或保管不当引起差错或造成损失；同时，说明书也是一种商业产品的广告宣传书，可以用来吸引顾客，扩大产品及企业的影响力。因此，产品说明书除了说明产品的特点，还要语言简练、流畅，通俗易懂，有时还要用一些专业术语。例如：

multi-functions　　功能多样　　　　　　fine workmanship　　做工精良
reliable and durable　　经久耐用　　　　easy operation　　操作方便
send image to application after saving　　存档后将影像传送到应用程序

二、电子产品说明书的组成

1. 电子产品说明书的封面与目录

封面（Cover）：通常印有实物照片或与产品有关的图案，并用生动的图像或鲜明的标志注明产品的名称。有的产品说明书在封面或封底标明生产单位、厂址、电话、邮编等，以便用户和厂家取得联系。

目录（Contents）：说明各章节名称及页码。

下面是联想系列微机说明书的封面与目录编排：

```
                    Lenovo（联想）

        联想系列微机
        技术手册
                    LXH—P1569
            Monitor User's Manual（显示器使用手册）
```

目录：（Contents）
中文　操作步骤　1 ……………………………………………………………… 21
English Operating Instructions　22 ……………………………………………… 42

2. 电子产品概述（Introduction）

本部分主要说明产品的主要功能特点、性能、操作程序以及注意事项等（有时把产品特色单列一项）。

下面是 Sports Network Walkman（运动型网络随身听）说明书的产品概述：

Introduction

The new Sports Network Walkman (NW-S4) player has a sleek pen shape and sports style, water-resistant design. It measures 5.2 in ×1 in and weighs 2.1 ounces. The 64 MB of embedded memory provides up to 120 minutes of skip-proof music and the device runs on one "AA" battery. The supplied arm strap makes the NW-S4 prefect for running, skateboarding and other active sports. Sports Network Walkman player also comes with a USB cable, headphones and one "AA" battery.

新运动型网络随身听（NW-S4）采用钢笔式平滑造型和运动风格，具有防水功能。其外观尺寸为5.2英寸×1英寸，重2.1盎司。64兆内存可存储长达120分钟的音乐（无跳读功能）。本产品由一节五号电池驱动。所提供的臂带可保证在跑步、滑板运动和从事其他体育活动中照常使用。该款产品还配有一根USB连线、一副耳机和一节5号电池。

3. 主要技术规格（Main Specifications）

本部分列出产品各项性能指标和产品工作条件的数据范围（如温度、电压、电流及变化范围等）。

下面是SHARP Grill Microwave Oven（夏普牌烧烤微波炉）说明书中技术规格的一部分：

AC Power Required：Microwave 1.00 kW（电力消耗：微波烹调时1.00千瓦）

Outside Power：Microwave 600 W （IEC-70 TEST PROCEDURE）

（输出功率：微波发生器600瓦（IEC-70测试方法））

Gill Heating Element 1.05 kW（烧烤发热器1.05千瓦）

Microwave Frequency：2 450 MHz（微波频率：2 450兆赫兹）

Outside Dimensions：464 mm（W）×300 mm（H）×362 mm（D）

（外形尺寸：464毫米（宽）×300毫米（高）×362毫米（深））

Cavity Dimensions：300 mm（W）×174 mm（H）×313 mm（D）

（炉箱内尺寸：300毫米（宽）×174毫米（高）×313毫米（深））

Weight：18.1 kg（重量：18.1千克）

4. 安全预防和维修保养（Safety Precautions and Maintenance）

本部分主要介绍产品的保养、常见的故障和处理方法。

下面是一则电子产品说明书中故障处理部分的部分内容：

Trouble-shooting（故障检修）

If you have any problem, read this manual again and check the countermeasure for the symptoms listed below. If the problem persists, contact your nearest authorized service center or dealer.

操作中若发生任何问题，请重新阅读本产品说明书，然后查看下面表格中所列举的问题解决方案。如果问题依然存在，请与离贵地最近的该产品授权服务中心或经销商联系。例如：

Good picture/Noisy sound	Check the TV SYSTEM setting
图像良好,但声音有杂音	查看 TV SYSTEM 键的位置设定正确与否

以上是电子产品英文使用说明书的一些特点,了解这些特点将有利于人们阅读并理解电子产品说明书的内容,从而根据说明书步骤更好地使用电子产品。

Reading Material

Basic Programming

Programming Hole Operation

The simplest operations to program are those related to producing holes. These include drilling, boring, taping and counter boring. The simplicity of programming lies in the fact that the programmer only needs to specify the coordinates of a hole center and the type of machine motions to be performed at the center. A fixed cycle, if used properly, takes over and causes the machine to execute the required movements. The controller stores a number of fixed cycles that can be recalled for use in programs when needed. This reduces the programming time length of tape required.

A fixed cycle is programmed by entering in one block information: the X and Y coordinates, the Z-axis reference plane (R), and the final Z-axis depth. In order to make it easier for you to understand the meaning of fixed cycle, let's see the G81 cycle in Figure 9-1.

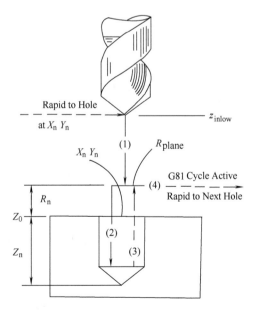

Figure 9-1 G81 Cycle

A G81 cycle causes the machine to:

(1) Rapid the tool from the Z position on the R.

(2) Drill the hole to a depth Z at federate F.

(3) Rapid back to either the R or Z position.

(4) Rapid to the center of the next hole if the X and Y coordinates of that hole programmed in the next block.

Programming Linear Profiles

Linear profiling involves cutting contours composed of straight lines only. The lines may be horizontal, vertical, or at any angle. We will use liner interpolation. Linear interpolation is used in part programming to make a straight cutting motion from the start position of the cut to its end position. Linear interpolation mode is designed for actual material removed, such as contouring, pocketing, face milling and many other cutting motions.

Three types of motion can be generated in the interpolation mode:

Horizontal motion, single axis only;

Vertical motion, single axis only;

Angular motion, multiple axes.

Let's see the linear profile milling though Figure 9-2.

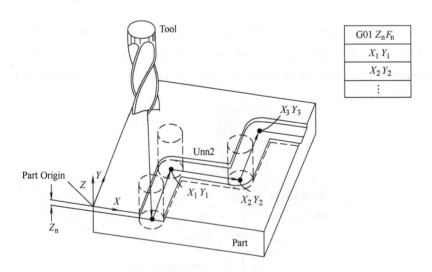

Figure 9-2 Linear Profile Milling

G01: Specifies the linear interpolation mode. The tool is moved at the programmed feed rate along a straight line.

Z_n: n specifies the absolute depth of the cut.

F_n: n specifies the feed rate of the tool into the material and along each subsequent straight line programmed. If not entered, the system will use the last feed rate programmed.

X_1Y_1: Specify the absolute coordinates of the cutter center at the end of line 1 cut, line 2 cut, and so on.

New Words and Phrases

  recall [riˈkɔːl] *vt.* 调用
  contouring [kənˈtuəriŋ] *n.* 轮廓
  pocketing [ˈpɔkitiŋ] *n.* 凹槽
  simplicity [simˈplisiti] *n.* 简单，朴素，直率
  linear [ˈliniə] *adj.* 直线的，线性的
  interpolation [inˌtəːpəuˈleiʃən] *n.* 插补
  counter boring 锪孔
  reference plane R 点平面，参考平面
  face milling 面铣削
  take over 接受，接管

Unit 10

Text

G-codes

G02 Circular Interpolation (CW, as show in Figure 10-1)

Format: N G02 X Y Z I J K F (I, J, K specify the radius)

Or: N G02 X Y Z R F (R specifics the radius)

Circular interpolation is more commonly known as radial (or arc) feed moves. The G02 command is used speciously for all clockwise radial feed moves, whether there are quadratic arcs, partial arcs, or complete circles, as long as they lie any one plane.[1] The G02 command is modal and is subject to a user definable feedrate.

Example: G02 X2 Y1 I0 J-1

The G02 command requires an endpoint and a radius in order to cut the arc. The start point is (X1, Y1) and the endpoint is (X2, Y2). To find the radius, simply measure the relative (or incremental) distance from the start point to the center point. This radius is written in terms of the X and Y distance. To avoid confusion, these are assigned variables called I and J, respectively.

Example: G02 X2 Y1 R1

You can also specify G02 by entering the X and Y endpoint and then R for the radius. An easy way to determine the radius values (I and J values) is by making a small chart:

Center point   X1   Y1
Start point   X1   Y2
Radius   I0   J-1

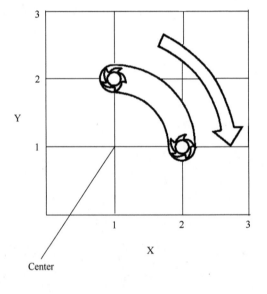

Figure 10-1 G02 Command

Finding the I and J values is easier than it first seems. Follow these steps:

1. Write the X and Y coordinates of the arcs center point.
2. Below these coordinates, write the X and Y coordinates of the arcs start point.
3. Draw a line below this to separate the two areas to perform the subtraction.

Result: G2 X2 Y1 I0 J-1 F5

4. To find the I value, calculate the difference between the arcs start point and center point in the X direction. In this case, both X values are 1. Hence there is no difference between them, so the I value is 0. To find the J value, calculate the difference between Y2 and Y1 is down 1 inch so the J value is -1.

G03 Circular Interpolation (CCW, as show in Figure 10-2)

Format: N G03 X Y Z I J K F   (I,J,K specify the radius)

Or: N G03 X Y Z R F   (R specifies the radius)

The G03 command is specifically for all counter clockwise radial feed moves, whether there are quadratic arcs, partial arcs, or complete circles, as long as they lie in any one plane. The G03 is modal and is subject to a user definable feedrate.

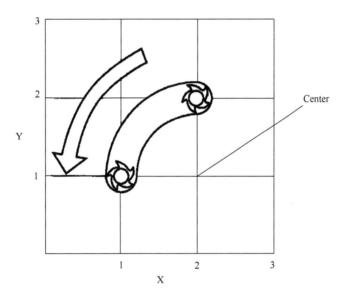

Figure 10-2 G03 Command

G04 Dwell

Format: N C04 P

G04 command is a non-modal command that halts all axis movement for a specified time, while the spindle continues revolving at the specified rpm. A Dwell is used largely in drilling operation, which allows for the clearance of chips. Use of the Dwell command is also common after a mill plunge move and prior to staring a linear profile move. This command requires a specified duration, denoted by the letter P, and followed by the time in seconds.

New Words and Phrases

confusion   [kən'fju:ʒən]   n. 混乱

chart [tʃɑːt] n. 图表
subtraction [səb'trækʃən] n. 减少
difference ['difərəns] n. 差异，差别
dwell [dwel] vi. 暂停
halt [hɔːlt] n. & v. 停止，中断
chip [tʃip] n. 碎片；筹码 vt. & vi. 碎裂
profile ['prəufail] n. 剖面，侧面，外形
duration [djuə'reiʃən] n. 持续时间，为期
circular interpolation 圆弧插补
clockwise（CW） 顺时针方向的
counter clockwise（CCW） 逆时针方向的
quadratic arcs 二次曲线
as long as 只要
in terms of 根据，按照

Notes

[1] The G02 command is used specifically for all clockwise radial feed moves, whether they are quadratic arcs, partial arcs, or complete circles, as long as they lie in any one plane.

G02 命令用于指定所有的顺时针圆弧插补运动，不管二次圆弧、部分圆弧还是整圆，只要它们在同一平面内，就可以用 G02 完成。

whether 引导状语从句，意思是"无论，不管……"；quadratic arcs, partial arcs, or complete circles 是三个并列词组，作状语从句的表语。

Translating Skills

科技英语翻译方法与技巧——电子产品英文使用说明书范例

电子产品英文使用说明书依据电子产品的具体特点有不同的写法，但无论采用何种写法，都应该突出产品的性能、特点、注意事项等。下面根据不同的内容，列举了一些实例，以便更好地帮助读者理解电子产品的英文使用说明书。

1. 产品说明的词句一般比较简单，常省去一些连词和冠词，大量使用祈使句和被动句语态。例如，数码摄像机使用说明书里是这样写的：

（1）Connecting the mains lead.（连接电源。）

1）Use the battery pack when using your camcorder outdoors.（室外使用摄像机，请用电池。）

2）Open the DC In jack cover.（打开直流电插座 IN 的盖子。）

3）Connect the plug with its ▲ make facing lens side.（接上插座，插头的 ▲ 符号朝向镜头方向。）

4）AC power adapter (supplied).（交流电源连接器［供电］。）

（2）Inserting a cassette.（放入录影带。）

1）Slide OPEN/EJECT in the direction of the arrow and open the lid.（沿箭头方向滑动 OPEN/EJECT 按钮，打开盖子。）

2）Push the middle portion of the back cassette to insert.（放入录像带时，推动录像带的后脊中部。）

3）Close the cassette compartment by pressing the PUSH mark on the cassette compartment.（推按录像带卡座上的 PUSH 按钮，关上录像带卡座。）

4）After the cassette compartment going down completely, close the lid until it clicks.（录像带卡座完全关上后，关上盖子，听到"咔嚓"声即可。）

（3）Recording a picture.（摄像。）

1）Remove the lens cap.（打开镜头盖。）

2）Set the POWER switch to CAMERA while pressing the small green button.（在按下绿色小按钮的同时，把电源开关 POWER 调制 CAMERA 的位置。）

3）To open the LCD panel, press OPEN. The picture appears on the LCD screen.（按下 OPEN 按钮，打开 LCD 显示屏，LCD 屏幕上出现图像。）

4）Press START/STOP. Your camcorder starts recording. To stop recording, press START/STOP again.（按下 START/STOP 按钮，摄像机开始摄像。停止摄像时，再按一次 ATART/STOP。）

（4）Monitoring the playback picture on the LCD screen.（在 LCD 屏幕上观看回放的图像。）

1）Set the POWER switch to VCR while pressing the small green button.（在按下绿色小按钮的同时，把电源开关 POWER 调至 VCR。）

2）Press NEW to rewind the tape.（按 NEW 键倒带。）

3）Press PLAY to start playback.（按 PLAY 键开始播放。）

2. 电子产品的主要技术规格（Main Specifications）需要列出产品各项性能指标和产品工作条件的数据范围等。下面是 Sports Network Walkman（NW-S4）说明书的技术规格：

Specifications（规格）

Compatibility Requirements（兼容性要求）：	WIN95/98/ME/2000
Memory Type（储存类型）：	Internal（内存）
LCD Display（液晶显示）：	Yes（是）
Product Line（产品系列）：	Network Walkman（网络随身听）
Audio Output Type（音频输出类型）：	Stereo（立体声）

Portable（便携式）： Yes（是）
Installed Memory（内存大小）： 64MB
Sound Output Mode（声音输出模式）： Stereo（立体声）
Product Color（产品颜色）： Black（黑色）
Supported Digital Audio Formats（所支持的数字音频格式）：MP3/WMA/AAC
Model（型号）：NW-S4
Operating System Compatibility（可兼容操作系统）：PC/Mac（PC/苹果机）
Connectivity Technology（接口类型）： USB（USB 接口）

3. 产品的安全预防和维修保养（Safety Precautions and Maintenance）对产品的安全使用和使用时间的长短起着非常重要的作用。

下面节选 Dell 产品说明书中产品保养的一些内容：

Do not operate the computer within a separate enclosure unless adequate intake and exhaust ventilation are provided on the enclosure that adhere to the guidelines listed above.

（计算机在另一个机箱内时，请勿进行操作，除非该机箱符合上面所列指导原则，即拥有足够的进气与排气的通风条件。）

Clean the display with a soft, clean cloth and water, Apply the water to the cloth; then stroke the cloth across the display in one direction, moving from the top of display to the bottom. Remove moisture from the display quickly and keep the display dry. Long-term exposure to moisture can damage the display. Do not use a commercial window cleaner to clean your display.

（显示器要用柔软干净的布和水清洗。用水将布沾湿，然后从显示器的上方往下，以同一个方向擦拭显示器。快速将水分从显示器上擦掉，以保持显示器干燥。显示器长时间暴露在水气中可能受到损害。请勿使用商用窗户清洗剂来擦拭显示器。）

If the computer doesn't start, or if you cannot identify the damaged components, contact Dell. (See your User's Guide or Owner's Manual for the appropriate contact information.)

如果计算机不启动，或您无法识别哪个组件受损坏，请联系 Dell。（请参阅您的《用户指南》或《用户手册》以获得合适的联系信息。）

以上是电子品英文说明说的一些实例，但产品说明书究竟需要哪几个部分以及怎样写，还要依据电子产品的具体需要而定。此外，电子产品英文说明应该选词准确、表达明了、行文流畅，为用户更好地使用电子产品带来方便。

Reading Material

M-codes

M-codes are miscellaneous functions that include actions necessary for machining but not those that are actual tool movements. That is, they are auxiliary functions, such as spindle on and off, tool changes, coolant on and off, program stops, and similar related functions. The following codes are described in more detail in the following sections.

M00   Program stop
M01   Optional program stop
M02   Program end
M03   Spindle on clockwise
M04   Spindle on counter clockwise
M05   Spindle stop
M06   Tool change
M08   Coolant on
M09   Coolant off
M30   Program end, reset to start
M98   Call subroutine command
M99   Return from subroutine command
Block Skip   Option to skip blocks that begin with "/"
Comments   Comments may be include in blocks with round brackets "( )"

M00   Program Stop
Format: N   M00

The M00 command is a temporary program stop function. When it is executed, all functions are temporarily stopped and will not restart unless and until prompted by user input.

The following screen prompt will be displayed with the CNCez simulators in: "Program Stop. Enter to Continue." The program will not resume unless and until Enter is pressed. The wording of this prompt varies by machine tool.

This command can be used in lengthy programs to stop the program in order to clear chips, take measurements, or adjust clamps, coolant hoses, etc.

M02   Program End
FORMAT: N   M02

The M02 command indicates an end of the main program cycle operation. Upon encountering the M02 command, the MCU switches off all machine operations (for example, spindle, coolant, all axes, and any auxiliaries) and terminates the program.

This command appears on the last line of the program.

M03   Spindle on Clockwise
Format: N   M03   S

The M03 command switches the spindle on in a clockwise rotation. The spindle speed is designated by the S letter address, followed by the spindle speed in revolutions per minute.

The spindle speed is shown during program simulation in the program status window. Its on/off status is shown in the system status window (CW, CCW, or OFF).

M04 Spindle on Counter Clockwise

Format: N M04 S

The M04 command switches the spindle on in a counterclockwise rotation. The spindle speed is designated by the S letter address, followed by the spindle speed in revolutions per minute.

The spindle speed is shown during program simulation in the program status window. Its on/off status is shown in the system status window (CW, CCW, or OFF).

M05 Spindle Stop

Format: N M05

The M05 command turns the spindle off. Although other M-codes turn off all functions (for example, M00 and M01), this command is dedicated to shutting the spindle off directly. The M05 command appears at the end of a program.

M30 Program End, Reset to Start

Formant: N M30

The M30 command indicates the end of the program data. In other words, no more program commands follow it. This is a remnant of the older NC machines, which could not differentiate between one program and the next, so an end of data command was developed. Now the M30 is used to end the program and reset it to the start.

Part Ⅵ CNC EDM

Unit 11

Text

Electric Discharge Machining

The use of thermoelectric source of energy in developing the nontraditional techniques has greatly helped in achieving an economic machining of the extremely low machinable materials and difficult jobs. [1] The process of material removal by a controlled erosion through a series of electric sparks, commonly known as electric discharge machining, was first started in the USSR around 1943. [2] Then onwards, research and development have brought this process to present level.

The basic scheme of electric discharge machining is illustrated in Figure 11-1.

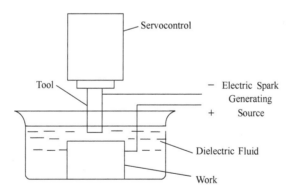

Figure 11-1 Basic Scheme of Electric Discharge Machining

When a discharge takes place between two points of the anode and the cathode, the intense heat generated near the zone melts and evaporates the material in the sparking zone. For improving the effectiveness, the workpiece and the tool are submerged in a dielectric fluid (hydrocarbon or mineral oils). It had been observed that if both the electrodes are made of the same materials, the electrode connected to the positive terminal generally erodes at a faster rate. For this reason, the workpiece is normally made the anode. A suitable gap, known as the spark gap, is maintained between the tool and the workpiece surfaces. The sparks are made to discharge at a high frequency with a suitable source. Since the spark occurs the spot where the tool and the workpiece surfaces are the closest and since the spot changes after each spark (because of the material removal after each spark), the spark travel all over the surface, and finally the work face conforms to the tool surface.[3] Thus, the tool produces the required impression in the workpiece. For maintaining the predetermined spark gap, a servocontrol unit is generally used. The gap is sensed through the average voltage across it and this voltage is compared with a present value.[4] The difference is used to control the servomotor. Sometimes, a stepper motor is used instead of a servomotor. Of course, for very primitive operations, a solenoid control is also possible, and with this, the machine becomes extremely inexpensive and simple to construct. Figure 11-2 schematically shows the arrangement of a solenoid controlled electric discharge machine. The spark frequency is normally in the range 200~500 000 Hz, the spark gap being of the order of 0.025~0.05 mm. The peak voltage across the gap is kept

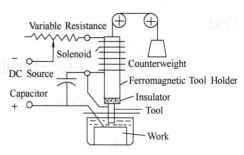

Figure 11-2 Solenoid Controlled Electric Discharge Machine

in the range 30~250 V. An mrr up to 300 mm^3/min[5] can be obtained with this process, the specific power being of the order of 10 W/(mm^3 · min). The efficiency and the accuracy of performance have been found to improve when a forced circulation of the dielectric fluid is provided. The most commonly used dielectric fluid is kerosene. The tool is generally made of brass or a copper alloy.

New Words and Phrases

thermoelectric [ˌθəːməuiˈlektrik] adj. 热电的
nontraditional [ˌnɔntrəˈdiʃənəl] adj. 非传统的，不符合传统的
erosion [iˈrəuʒən] n. 腐蚀，侵蚀
onwards [ˈɔnwəːdz] adv. 向前地，在先地
scheme [skiːm] n. 配置，图解；计划，设计，图谋，策划
anode [ˈænəud] n. 极，正极
cathode [ˈkæθəud] n. 阴极
evaporate [iˈvæpəreit] vt. (使) 蒸发，消失
effectiveness [iˈfektivnis] n. 效力
submerge [səbˈməːdʒ] vt. 浸没，淹没，湮没 vi. 潜水
dielectric [ˌdaiiˈlektrik] n. 电介质，绝缘体 a. 非传导性的
hydrocarbon [ˈhaidrəuˈkɑːbən] n. 烃，碳氢化合物

conform　　[kən'fɔːm]　　vt. 使一致，使顺从；符合，相似，吻合
impression　　[im'preʃən]　　n. 印象，感想；盖印，压痕
servocontrol　　['seːvəkən'trəul]　　n. 伺服控制，随动控制
sense　　[sens]　　n. 感觉，判断力，理性感；理解，认识
servomotor　　['səːvəuˌməutə]　　n. 伺服电动机，补助电动机
stepper　　['stepə]　　n. 步进者
solenoid　　['səulinɔid]　　n. 螺线管
schematically　　[skiˈmætikli]　　adv. 示意性地
order　　['ɔːdə]　　n. 次序，顺序；命令；定购　vt. 定购，定制
kerosene　　['kerəsiːn]　　n. 煤油，火油
brass　　[brɑːs]　　n. 黄铜，黄铜制品；厚脸皮
spark gap　　火花隙

Notes

[1] The use of thermoelectric source of energy in developing the nontraditional techniques has greatly helped in achieving an economic machining of the extremely low machinable materials and difficult jobs.

利用电热能开发非传统（加工）技术，有助于提高切削性能差、难加工材料的加工经济性。

[2] The process of material removal by a controlled erosion through a series of electric sparks, commonly known as electric discharge machining, was first started in the USSR around 1943.

通过控制一系列电火花的腐蚀来除去材料的工艺通常称为电火花加工，该工艺始于1943年的前苏联。

[3] Since the spark occurs the spot where the tool and the workpiece surfaces are the closest and since the spot changes after each spark (because of the material removal after each spark), the spark travel all over the surface, and finally the work face conforms to the tool surface.

由于火花产生在刀具和工件的最接近点，且火花产生后最接近点发生变化（因为产生火花后材料被切除），所以电火花移动经过整个平面，最后，工件的表面（廓形）与刀具相吻合。

[4] The gap is sensed through the average voltage across it and this voltage is compared with a present value.

火花隙能感知穿过它的平均电流并将该值与其现行值比较。

[5] An mrr up to 300 mm^3/min...

高达每分钟300立方毫米的金属切除量……　　mrr = metal removal rate。

Translating Skills

科技英语翻译方法与技巧——长难句的翻译

一、次序的处理

1. 采用顺译法

在科技英语中,有些句子虽较长,但它们叙述的内容基本上是按照所发生的时间先后安排,或者内容是按照逻辑关系安排,与汉语表达方式比较一致。因此可以按照原文的顺序译出。这种长句拆译称为"顺译法"。例如:

If, when an SCR is biased in the forward direction a current pulse is injected into the P region of junction 2 by momentarily applying voltage to the gate, junction 2 becomes forward-biased and conduction through the SCR begins.

当晶闸管正向偏置时,如果对控制极加了瞬时正电压,从而由一个脉冲流进入结2的P区,这时结2正向偏置,晶闸管开始导通。

However, fiber systems can carry so many more telephone conversations at the same time than wire pairs, and can carry them so much farther without amplification of regeneration, that when there are many telephone calls to be carried between points such as switching offices fiber systems arc economically attractive.

然而,光纤系统同时可容纳比线对多得多的电话线路,并且在不需要放大的情况下把信号送得更远。当两点之间,比如中继局有很多电话呼叫要传送的时候,光纤系统就是非常经济的。

2. 采用倒译法

"顺译法"在长句的翻译中实际上运用的较少,当英语句子中的逻辑次序和汉语相反,如表示条件、时间、让步等意义的从句位于主句之后时,往往需要采用另一种译法——"倒译法",把从句提到主句之前来翻译。例如:

A student of mathematics must become familiar with all the signs and symbols commonly used in mathematics and gear them in mind firmly, and be well versed in the definitions, formulas as well as the technical terms in the field of mathematics, in order that he may be able to build up the foundation of the mathematical subject and master it well for pursuing advanced study.

为了打好数学基础,掌握好数学以便学好深造,一个学数学的人必须和牢记数学中所有常用的符号,通晓定义、公式以及术语。

这是由一个主句(主语及三个并列的谓语)及一个从句(一个主语及两个谓语)组成的复合句。按照英语句子结构的习惯先表明看法及态度,再列举情况。这与汉语的表达习惯有所不同。汉语重"意合",句子内部以及上下之间多靠语义来连接;句子成分的先后位置通常是按时间顺序或逻辑顺序来安排的。所以必须从原文的后部开始翻译，逆着原文的顺序翻译。

This is why the hot water system in a furnace will operate without the use of a water pump, if the pipes are arranged so that the hottest water rises while the coldest water runs down again to the furnace.

如果把管子装成能让最热的水上升，最冷的水流回到锅炉去，那么锅炉中热水系统不用水泵就能运转，其道理就在于此。

二、结构的处理

1. 翻译采用短句

有些长难句或介词短语等修饰词，与主句的关系不十分密切，可遵照汉语多用短句的习惯，将它们译成短句分开叙述。例如：

The laser, its creation being thought to be one of today's wonders, is nothing more than a light that differs from ordinary lights only in that it is many times more intense and so can be applied in the field that no ordinary light has ever been able to penetrate into。

激光的发明虽然被认为是当代的一个奇迹，其实它也就是一种光。这种光不同于普通的光是因为它比普通的光强烈许多倍，因而能够应用于普通光无法穿透的地方。

2. 压缩句子，用词或短语代替

将某些从句压缩成一个词或词组，使句子更为精炼。例如：

With radar, we can "see" thing at a great distance and it shows us how far away they are, in which direction they lie, and whet movements they are making。

有了雷达，我们就能"看到"远方的物体，雷达还能向我们指明这些物体的距离、方位和运动方式。

For use where small size, light weight and portability are highly desirable, transistorized oscilloscopes are preferred over vacuum-tube types。

在应用中如果需要体积小、重量轻以及携带方便，那么采用晶体管示波器比采用电子管优越。

三、综合法

有些英语长难句结构复杂，表达方法与汉语差异很大，这时需要仔细推敲原文，按汉语习惯进行综合处理。例如：

The efforts that have been made to explain optical phenomena by means of the hypothesis of a medium having the same physical character as an elastic solid body led, in the first instance, to the understanding of a concrete example of a medium which can transmit transverse vibrations, and at a later stage to the definite conclusion that there is no luminiferous medium having the physical character assumed in the hypothesis。

为了揭示光学现象，人们曾试图假定有一种具有与弹性固体相同的物体的介质。这种尝试的结果，最初曾使人们了解到一种能传输横向振动的具有上述假定的那种物理性质的发光介质。

Reading Material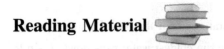

Laser Beam Machining

Like a beam of high velocity electrons, a laser beam is also capable of producing very high power density. Laser is a highly coherent (in space and time) beam of electromagnetic radiation with wavelength varying from 0.1~70μm. However, the power requirement for a machining operation restricts the effectively usable wavelength range to 0.4~0.6μm. Because of the fact that the rays of a laser beam are perfectly parallel and monochromatic, it can be focused to a very small diameter and can produce a power density as high as 107 W/mm^2.

For developing a high power, normally a pulsed ruby laser is used. The continuous CO_2-N_2 laser also been successfully used in machining operations.

Figure 11-3 shows a typical pulsed ruby laser. A coiled xenon flash tube is placed around the ruby rod and the internal surface of the container walls is made highly reflecting so that maximum light falls on the ruby rod for the pumping operation. The capacitor is charged and a very high voltage is applied to the triggering electrode for initiation for the flash. The emitted laser beam is focused by a lens system and the focused beam meets the work surface, removing a small portion of the material by vaporization and high speed ablation. A very small fraction of the molten metal

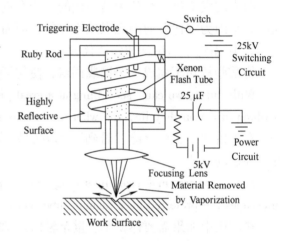

Figure 11-3　Schematic Diagram of Laser Beam Machining (LBM)

is vaporized so quickly that a substantial mechanical impulse is generated, throwing out a large portion of the liquid metal. Since the energy release by the flash tube is much more than the energy emitted by the laser head in the form of a laser beam, the system must be properly cooled.

The efficiency of the LBM process is very low, about 0.3%~9.5%. The typical output energy of a laser is 20 J with a pulse duration of 1 millisecond. The peak power reaches a value 20 000W. The divergence of the beam is around 2×10^{-5} Gy, and, using a lens with a focal length of 25 mm, the spot diameter becomes about 50μm.

Like the electron beam, the laser beam is also used for drilling micro holes and cutting very narrow slots. Holes up to 250μm diameter can be easily drilled by a laser. The dimensional accuracy is around 0.025mm. When the workpiece thickness is more than 0.25mm, a taper of

0.05mm per millimeter is noticed.

New Words and Phrases

beam	[biːm]	n. 梁，桁条，横梁；（光线的）束，电波 vt. 播送	
coherent	[kəuˈhiərənt]	adj. 粘在一起的，一致的，连贯的	
radiation	[ˌreidiˈeiʃən]	n. 发散，发光，发热，辐射，放射线，放射物	
wavelength	[ˈweivˌleŋθ]	n. 波长	
monochromatic	[ˈmɔnəukrəuˈmætik]	adj. 单色的，单频的	
xenon	[ˈzenɔn]	n. 氙	
flash	[flæʃ]	n. 闪光，闪现，一瞬 vt. 反射 adj. 火速的	
rod	[rɔd]	n. 杆，棒	
trigger	[ˈtrigə]	vt. 引发，引起，触发 n. 扳机	
lens	[lenz]	n. 透镜，镜头	
ablation	[æbˈleiʃən]	n. 消融，切除	
vaporize	[ˈveipəraiz]	vt. （使）蒸发，汽化	
substantial	[səbˈstænʃəl]	adj. 坚固的；实质的，真实的；充实的	
emit	[iˈmit]	vt. 发出，放射；吐露，散发；发表，发行	
divergence	[daiˈvəːdʒəns, di-]	n. 分离，散开；分歧	
micro	[ˈmaikrəu]	adj. 微小的	
thickness	[ˈθiknis]	n. 厚度，浓度；稠密	

Unit 12

Text

Wire-Cut EDM (1)

Principle of EDM

Electrical discharge machining, commonly known as EDM, is a controlled metal removal process whereby an electric spark is used to cut (erode) the workpiece, which then takes the shape opposite to that of the cutting tool or electrode.[1] The electrode and the workpiece are both submerged in a dielectric fluid, which is generally light lubricating oil. This dielectric fluid should be a nonconductor of electricity. A servomechanism maintains a gap of about 0.0005 to 0.001 in. (0.012 to 0.025mm) between the electrode and the work, preventing them from coming into contact with each other. A direct current of low voltage and high amperage is delivered to the electrode at the rate of approximately 20 000 hertz (Hz). These electrical energy impulses become sparks which jump the gap between the electrode and the workpiece through the dielectric fluid.[2] Intense heat is created in the localized area of the spark impact; the metal melts and a small particle of molten metal is expelled from the workpiece. The dielectric fluid, which is constantly being circulated, carries away the eroded particles of metal and also helps in dissipating the heat caused by the spark.

There are two types of EDM machines used in industry: the vertical EDM machine and the wire-cut EDM machine (Figure 12-1 a, b). Since the wire-cut EDM is generally used for machining complex forms which require NC programming, only this type will be discussed in detail.

Wire-Cut EDM

The wire-cut EDM is a discharge machining which uses NC movement to produce the desired contour or shape on a part. It does not require a special-shaped electrode; instead, it uses a continuous traveling wire under tension as the electrode. The electrode or cutting wire can be made of brass, copper, or any other electrically conductive material ranging in diameter from 0.002 to 0.012 in (0.05 to 0.30 mm). The path that the wire follows is controlled along a two-axis (XY) contour, eroding (cutting) a narrow slot, through the workpiece.[3] This controlled movement is continuous and simultaneous in increments of 0.000 05 in (0.001 2 mm). Any contour may be cut to a high degree of accuracy and is repeatable for any number of successive parts. A dielectric fluid, usually deionized water which is constantly being circulated, carries away the eroded particles of metal.[4] The dielectric fluid maintains the proper conductivity between the wire and the workpiece and assists in reducing the heat caused by the spark.

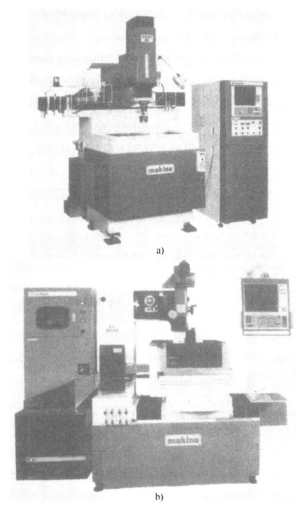

Figure 12-1 Electrical Discharge Machines
a) Vertical EDM Machine   b) Wire-cut EDM Machine

Parts of the Wire-Cut EDM Machines

The main parts of the wire-cut EDM Machines are the bed, saddle, table, column, arm, UV axis head, wire feed and dielectric systems, and machine control unit (MCU) (Figure 12-2).

Bed: The bed is a heavy, rugged casting used to support the working parts of the wire-cut EDM Machines. Guide rails ("way") are machined on the top section and these guide and align major parts of the machine.

Saddle: The saddle is fitted on top of the guide rail and may be moved in the XY direction by feed servomotors and ball lead screws.

Table: The table, mounted on top of the saddle, is U-shaped and contains a series of drilled and tapped holes equally spaced around the top surface. These are used for the workpiece holding and clamping devices.

Column: The column, which is attached to the bed, supports the wire feed system, the UV axes, and the capacitor switches.

Wire feed system: The wire feed system is used to provide a continual feed of new wire (electrode) for the cutting operation. The wire is fed from a supply spool through a series of guides mid guide rollers which apply tension to the wire.[5] The wire travels in a continuous path past the workpiece, and the used wire is rewound on a take-up spool.

Dielectric system: The dielectric system contains filters, an ion exchanger, and a cooler. This system provides a continuous flow of clean deionized water at a constant temperature. The deionized water stabilizes the cutting operation, flushes away particles of electrode and workpiece material that have been eroded, and cools the workpiece.

MCU: The MCU can be separated as three individual panels.

1) The control panel for setting the cutting conditions.

2) The control panel for machine setup.

3) The control panel for manual data input (MDI) and cathode-ray tube (CRT) character display (Figure 12-3).

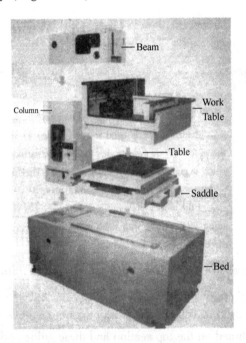

Figure 12-2   The Main Parts of a Wire-cut EDM Machine

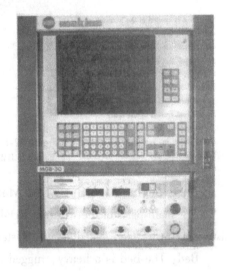

Figure 12-3   Control Panel

New Words and Phrases

spark   [spɑːk]   n.   火花

erode   [iˈrəud]   v.   腐蚀，电蚀

submerge   [səb'mɜːdʒ]   v. 淹没，浸入
nonconductor   [ˈnɔnkənˈdʌktə]   n. 绝缘体
amperage   [ˈæmpɛəridʒ]   n. 安培数
impulse   [ˈimpʌls]   n. 脉冲
discharge   [disˈtʃɑːdʒ]   v. 放电
dissipate   [ˈdisipeit]   v. 驱散；耗散
dielectric   [ˌdaiiˈlektrik]   n. 电介质；绝缘体
deionize   [diːˈaiəˌnaiz]   v. 除去离子
rugged   [ˈrʌgid]   adj. 结实的，坚固的
align   [əˈlain]   v. 调整，校准
clamp   [klæmp]   v. 夹住
ion   [ˈaiən]   n. 离子
capacitor   [kəˈpæsitə]   n. 电容器
expel   [iksˈpel]   v. 驱除，开除；排出
cathode-ray tube（CRT）   阴极射线管
wire-cut EDM   线切割电火花加工
guide roller   导轮

Notes

[1] Electrical discharge machining, commonly known as EDM, is a controlled metal removal process whereby an electric spark is used to cut (erode) the workpiece, which then takes the shape opposite to that of the cutting tool or electrode.

电火花加工通常简称 EDM，它是采用电火花放电腐蚀工件来控制金属切削的加工方法。这样加工出的工件形状与切割刀具或电极的形状相反。

句中关系副词 whereby 相当于 by which, by what，意思是"借以"；which 引出的非限定性定语从句修饰 the workpiece。

[2] These electrical energy impulses become sparks which jump the gap between the electrode and the workpiece through the dielectric fluid.

脉冲电能在电极与工件间的工作液中火花放电。

句中 which 引导的定语从句修饰 sparks, dielectric fluid 可译为"工作液"。

[3] The path that the wire follows is controlled along a two-axis (XY) contour, eroding (cutting) a narrow slot through the workpiece.

切割线被控制在 X-Y 平面内沿工件轮廓曲线移动，从而在工件上电蚀出一条狭缝。

句中 that the wire follows 为定语从句，修饰 The path；句子的主干是 The path is controlled along；eroding... 是现在分词短语作伴随状语。

[4] A dielectric fluid, usually deionized water which is constantly being circulated, carries

away the eroded particles of metal.

工作液通常是去除离子的水,它不断循环流动,冲走熔蚀的金属微粒。

deionized water 去离子水;which is constantly being circulated 为定语从句,修饰 deionized water;句子主干是 A dielectric fluid carries away the eroded particles of metal.

[5] The wire is fed from a supply spool through a series of guides and guide rollers which apply tension to the wire.

新线从贮丝筒的供线端绕过一系列导轮与导柱,这些导轮和导柱给切割线提供一定张紧力。

which 引导的定语从句修饰 guides and guide rollers,说明它所起的作用。supply spool 贮丝筒;apply… to… 把……施加于……。

Translating Skills

科技英语翻译方法与技巧——英语否定式的翻译

英语和汉语在表达形式上都有肯定和否定形式之分,翻译时,一般应把肯定形式译成肯定形式,否定形式译成否定形式。但有时两种语言对否定概念的表达上有很大差别,因此在翻译过程中,有时要做否定语气的转换,有时甚至要做反面处理,即英语中否定形式有时可译成汉语中的肯定形式,英语中的肯定形式反而可译成汉语中的否定形式。翻译时要注重两种语言表达否定概念时在词汇语法以及语言逻辑方面的差别。

例如:

The earth does not move round in the empty space.

地球不是在空无一物的空间中转动。(不是"地球在空无一物的空间中不转动。")

All metals are not good conductors.

并非所有的金属都是良导体。(不是"所有金属都不是良导体。")

可见,英语中的否定形式与否定概念并不都是一致的。第一句中,形式上是一般否定(谓语否定),但实际上却特指否定(其他成分的否定);第二句中,似乎是全部否定,实际上却是部分否定。

一、否定成分的转译

有些英语的否定句虽然用一般否定(谓语否定)的形式,但在意义上却是特指否定,即对其他成分的否定。反之,有些句子形式上是特指否定,而意义上却是一般否定。翻译时应按汉语习惯进行否定成分的转译。例如:

Liquids have no definite shapes.

液体没有固定的形式。(原文否定宾语,译文否定谓语,汉语中不习惯说"液体有未定的固定形式。")

We don't think that is possible.

我们认为这是不可能的。（原文否定主句谓语，译文否定从句谓语，汉语不习惯说"我们不认为这是不可能的。"）

应该注意，英语中表达对某一问题持否定见解的句子，其否定词往往放在谓语动词（think, believe, consider, suppose 等）之前，按汉语习惯，不能译为"不认为……"、"不觉得……"，而应译为"认为……不"、"觉得……不"。

This version is not placed first because the fuel was finished.

这个方案并不因为简单而放在首位。（原文否定谓语，译文否定状语 because，不能译为"这个方案因为简单而没有放在首位。"）

The engine did not stop because the fuel was finished.

发动机不是因为燃料用完而停止的。（原文否定谓语，译文否定原因状语，不能译为"发动机没有停止，因为燃料用完了。"）

The mountain is not valued because it is high.

山的价值并不在于它的高。（原文否定谓语，译文否定原因状语。不能译为"山因其高而无价值。"）

注意："not... because"是一种用法特殊的否定结构。连词 because 与否定词 not 连用时，可能有两种不同的含义，因为 not 有时否定 because（译为"不因为"），如 Pure iron is not used in industry because it is too soft. 译为"纯铁因为太软而不用在工业上。"翻译时必须从上下文的逻辑意义上判断。

二、否定语气的改变

由于英语与汉语表达否定概念上的不同，翻译时有时要把英语中的否定句译成汉语的肯定句，才符合汉语的表达习惯，这种译法称为"反面着笔"。

1. 否定句译成肯定句

Metals do not melt until heated to a definite temperature.

金属要加热到一定的温度才会融化。（不译为"金属不会融化直到加热到一定的温度为止。"）

An insulator is a substance that contains very few free electrons and whose atoms have on loosely held orbital electrons.

绝缘体是一种含有很少的自由电子的材料，它的原子都把电子紧紧地吸在轨道上。（不译为"……，它的原子不具有被松散地吸在轨道上的电子。"）

Ball bearings are precision-made bearing that make use of the principle that "nothing rolls like a ball".

滚珠轴承是精密的轴承，采用"球形最善于滚动"的原理。

2. 肯定句译成否定句

"反面着笔"的另一种译法是将英语的肯定句译成汉语的否定句，以使译文符合汉语的表达习惯。例如：

Worm gear drives are quiet, vibration free and extremely compact.

蜗轮传动没有噪音，没有振动，而且非常紧凑。（不译为"蜗轮传动是平静的，无振动的，而且是非常紧凑的。"）

The influence of temperature on the conductivity of metals is slight.

温度对金属导电性的影响不大。（不能译为"……是轻微的。"）

The structure will prove weak in service.

在使用中会证明该物体是不牢固的。（不能译为"……是弱的。"）

Reading Material

Wire-Cut EDM (2)

Various systems on wire-cut EDM machines play an important part in the efficient operation of the machine tool. Servo systems, dielectric fluid, electrode, and the MCU are the main operating components of wire-cut EDM machines.

The Servomechanism

The EDM power supply controls the cutting current levels and the feed rate of the drive motors. It also controls the travelling speed of the wire. EDM machines are equipped with a servo control mechanism that automatically maintains a constant gap of approximately 0.001 to 0.002 in (0.025 to 0.05 mm) between the wire and the workpiece. It is important that there be no physical contact between the wire (electrode) and the workpiece; otherwise arcing could damage the workpiece and break the wire. The servomechanism also advances the wire into the workpiece as the operation progresses and senses the work-wire spacing and slows or speeds up the drive motors as required to maintain the proper arc gap.

Precise control of the gap is essential to a successful machining operation. If the gap is too large, ionization of the dielectric fluid does not occur and machining can not take place. If the gap is too small, the wire will touch the workpiece, causing it to melt and break.

Precise gap control is accomplished by a circuit in the power supply comparing the average gap voltage to a preselected reference voltage. The difference between the two voltages is the input signal, which tells the servomechanism how far and how fast to feed the wire and when to retract from the workpiece. This is usually indicated by a reverse servo light.

When chips in the spark gap reduce the voltage below a critical level, the servomechanism causes the wire to retract until the chips are flushed out by the dielectric fluid. The servo system should not be too sensitive to "short-lived" voltages caused by chips being flushed out; otherwise the wire would be constantly retracting, thereby seriously affecting machining rates.

The Dielectric Fluid

One of the most important factors in a successful EDM operation is the removal of the particles (chips) from the working gap. Flushing these particles out of the gap with the dielectric fluid will produce good cutting conditions, while poor flushing will cause erratic cutting and poor machining conditions.

The dielectric fluid in the wire-cut EDM process is usually deionized water. This is tap water that is circulated through an ion exchange resin. The deionized water makes a good insulator, while untreated water is a conductor and is not suitable for the EDM process. The ion exchanger is a compound of positive ion exchange resin (cation) and negative ion exchange resin (anion). When water is applied to this compound, the ion exchange reaction starts and is continuously repeated until all the impurities are completely removed from the water, thus producing pure water。

The amount of deionization is measured by its specific resistance. For most operations, the lower the resistance, the faster the cutting speed. However, the resistance of the dielectric fluid should be much higher when carbides and high-density graphite are cut (Figure 12-4).

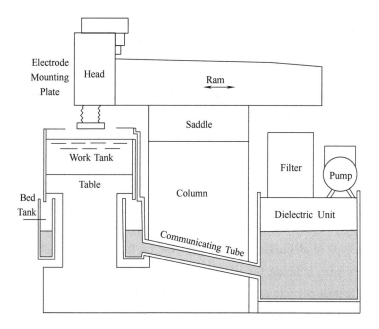

Figure 12-4 The Dielectric Fluid System

The dielectric fluid in the wire-cut EDM process serves several functions:
1) It helps to initiate the spark between the wire (electrode) and the workpiece.
2) It serves as an insulator between the wire and the workpiece.
3) It flushes away the particles of disintegrated wire and workpiece to prevent shorting.
4) It acts as a coolant for both the wire and the workpiece.

The dielectric fluid must be circulated under constant pressure if it is to flush away the particles and assist in the machining process. The flow of water is governed by two control valves: one controlling the water flowing above the workpiece and one controlling the flow under the workpiece. When starting to supply tile water for the cutting process, apply a steady amount "down-stream" and then slowly apply the water from underneath until there is a flaring effect at the top surface of the workpiece (Figure 12-5).

Figure 12-5　A Steady Stream of Dielectric Must be Supplied to the Workpiece

If red sparks occur during the cutting operation, the water supply is inadequate. To overcome this problem, increase the flow of water until blue sparks appear. Note: Excessive water flow can cause the wire to be deflected, causing erratic cutting and losing machining accuracy.

The Electrode

The electrode in wire-cut EDM is a spool of wire ranging from 0.002 to 0.012 in (0.05 to 0.30 mm) in diameter and from 2 to 100 lb (0.90 to 45.36 kg) in weight. The length of wire on a spool can provide over 500 h of unattended machining time. The electrode of the wire-cut machine continuously travels from a supply spool to a take-up spool so that it is constantly renewed. When this type of electrode is used, the wear on the wire does not affect the accuracy of the cut because new wire is being fed past the workpiece continuously at rates from a fraction of an inch to several inches per minute. Both the electrode wire and the material removal rate from the workpiece depend on such things as the material's thermal conductivity, its melting point, and the duration and intensity of the electrical pulses. As in conventional machining, some EDM materials have better cutting and wearing qualities than others; therefore, electrode materials must have the following characteristics:

1) Be a good conductor of electricity;
2) Have a high melting point;
3) Have a high tensile strength;
4) Have good thermal conductivity;
5) Produce efficient metal removal from the workpiece.

Research and experimentation are continually increasing metal removal rates and finding good, economical materials for the manufacture of wire-cut EDM electrodes. Brass, copper, tungsten, molybdenum, and zinc are some of the materials which have found certain applications as electrode materials. The most widely used materials brass wire 0.008 in (0.20 mm) in diameter. A normal overcut of about 0.001 in (0.025 mm) per side will produce an internal comer with a radius of about 0.005 in (0.12 mm). Smaller diameters of brass wire are also used for many applications. Tungsten and molybdenum wire, which have a very high melting point and high tensile strength, permit small-diameter wire of 0.002 in (0.05 mm) in diameter to be used for cutting fine radii and intricate shapes.

Stratified wire, which consists of a copper core with a thin surface layer of zinc, gives the wire the high conductivity of copper combined with the cooling effect of zinc. This allows the use of higher current, increasing the energy level of the EDM spark and, as a result, the metal removal rate.

New Words and Phrases

servo	[ˈsəːvəu]	n.	伺服系统
ionization	[ˌaiənaiˈzeiʃən]	n.	离子化，电离
retract	[riˈtrækt]	v.	缩回，回退；取消
erratic	[iˈrætik]	adj.	不稳定的
cation	[ˈkætaiən]	n.	阳离子
anion	[ˈænaiən]	n.	阴离子
resin	[ˈrezin]	n.	树脂
graphite	[ˈgræfait]	n.	石墨
initiate	[iˈniʃieit]	v.	开始；发动
spool	[spuːl]	n.	线轴
brass	[brɑːs]	n.	黄铜
tungsten	[ˈtʌŋstən]	n.	钨
molybdenum	[məˈlibdinəm]	n.	钼
zinc	[ziŋk]	n.	锌
stratified	[ˈstrætifaid]	adj.	分层的
ion exchanger			离子交换剂

Appendix 1　参考译文

第一部分　数控加工介绍

第 1 单元

课文　　　　　　　　　　　　计算机数控加工

计算机数控的定义及其组成

计算机数控机床就是在 NC 机床的基础上附加一个内置式计算机，这种内置式计算机通常称为机床控制单元或 MCU。NC 机床的控制单元通常由硬件构成，也就是说机床的功能是由控制器内置的物理电子元器件控制的。简单地说，数字控制是基于代码自动操作制造类机床的一种方法，这些代码由字母、数字和特殊字符组成。所谓的程序就是用于执行某一操作的一整套代码表示的指令，而这些程序是存放在计算机的随机存储器（RAM）中的，控制器可以对它们进行重新调整、编写以及处理等操作。因此，可以通过对计算机进行编码从而在加工时实现机床操作的各种功能。

CNC 系统中常见的组成部分如图 1-1 所示。

机床控制单元：生成、存储和处理 CNC 程序。机床控制单元还包括机床运动控制器，该控制器是一个可执行的软件程序。

NC 机床：响应来自机床控制单元的程序信号并对零件进行加工。

CNC 设备的类型

加工中心是 CNC 技术中最前沿的发展。这些系统装有自动换刀装置，可换刀具数高达 90 把甚至更多。许多加工中心还装有可移动的长方形工作台，即托盘系统。这种托盘系统用来自动装卸工件。加工中心只需一次装卡就可以完成铣、钻、攻螺纹、镗等多种加工操作。此外，一些加工中心利用分度头可以对零件不同的面和特定的角度进行加工。通过减少零件在机床间转移，加工中心能够节省生产时间与成本。

具有大容量刀库的车削中心同样在现代制造厂中占据着显著位置。这些 CNC 机床能对旋转的工件同时进行多种车削加工。

除加工中心和车削中心之外，CNC 技术还被广泛用于其他许多类型的设备，其中包括线切割电火花加工机床和激光切割机床。

线切割电火花加工机床用一根很细的切割丝做电极。切割丝在两个导轮之间拉紧，像带锯一样切割工件。切割丝水平移动所产生的电火花会将材料腐蚀并去除，而 CNC 就用来控制水平工作台的移动。

激光切割CNC机床能利用一束强烈的聚焦光线来切割工件。材料在光线的照射下温度会骤然上升并汽化。若光束的能量足够高，就会将材料穿透。由于没有机械切削力，所以激光加工的工件变形非常小，激光切割机对加工窄槽和钻孔非常有效。

阅读材料　　　　　　　　　　**数控的历史**

欢迎来到数控世界。由于数字控制（NC）使制造系统柔性更强，因而已广泛用于工厂与车间。简而言之，数控机床是通过代码指令使机床沿着预编程轨迹自动定位的机床，这里的关键词是"预编程"与"代码化"。机床运行前，必须有人确定需要机床执行什么操作，并将此信息译成数控装置能够识别的代码。换句话说，必须有人对机床编程。

人们可以手工编写加工程序，称为手工编程；亦可借助计算机对机床编程，称为计算机辅助编程（CAP）。手工编好的程序有时可从机床键盘送入控制器，这叫做手动数据输入（MDI）。

微电子技术与微型计算机的发展已使计算机可用于现代数控机床的控制单位，取代了早期NC机床的读带机。换句话说，程序是由机床上的计算机存取和执行的，而不再直接取自穿孔带。这种计算机数控（CNC）机床就是当今制造的数控（NC）机床。

1947年，Parsons公司的John Parsons着手进行一项试验，他想用三轴曲率数据控制机床运动以生产飞机零件。1949年，Parsons公司与美国空军签订了制造第一台数控机床的合同。1951年，麻省理工学院承担了这一项目。1952年，MIT（麻省理工学院）使用实验室制造的控制器和辛辛那提立式主轴展示三轴联动获得成功，这标志着数控时代的到来。到了1955年，几经改进之后，数控技术开始应用于工业生产。

早期的NC机床能运行穿孔卡与穿孔带，二者中以穿孔带更为通用。但是，鉴于更换、编辑纸带费时费力，后来便采用计算机作为编程的辅助工具。计算机的引入有两种形式：一是计算机辅助编程语言，二是直接数字控制（DNC）。有了计算机辅助编程语言，程序员可用一套通用"混杂英语"命令编写NC程序，然后由计算机将其释译为机器码并制成穿孔带。直接数字控制（如图1-2所示）是指用一台计算机对一台或多台数控机床实施部分或整体控制。虽然有些公司运用DNC已获得成功，但是，扩大计算机容量、购买软件、协调DNC系统等花费使这种系统并不适合所有公司，而只适用于一些大公司。

最近，一种叫做分布式数字控制的新型DNC系统已经开发出来（如图1-3所示）。它用计算机网络来协调多台DNC机床的运行。这种方式最终有可能用来协调整个工厂的运转。分布式数字控制解决了在协调直接数字控制系统时存在的一些问题。另外还有一种作为前面介绍过的系统的副产品的分布式数字控制系统，其整个NC程序可从主机直接传输到机床控制器。另外，该系统可将程序从主机传输到车间的个人计算机（PC）中并保存直到需要时，然后再传输到机床控制器。

第 2 单元

课文 **数控的优点**

　　近期的研究表明，在加工零件时，只有一少部分时间真正用于加工工件。我们假定把一个零件从运达工厂时的毛坯或棒料加工为成品需要 50 小时。在这段时间里，工件在机床上的时间只有 2.5 小时，用于切削的时间仅有 0.75 小时，其余时间都用在等待加工、运输、安装、装夹、拆卸、检验、设置转速与进给、换刀上。

　　通过转速与进给量的选择、刀具在待切削表面间的快速移动、自动夹具的运用、刀具的自动切换、切削液的控制、在线检测及零件的自动装卸，NC 缩短了非切削加工时间。上述因素，加之无需再培训机床操作工，使整个加工过程耗时大量减少，促进了 NC 的推广应用。NC 的主要优点如下：

　　1）机床可以自动运行，亦可半自动运行，使用时按需选取。

　　2）零件的柔性制造更为简便。更换纸带即可加工不同零件。

　　3）减小存储空间。通常采用简单的工装夹具，因而减少了存储夹具的数量。

　　4）加工小批量零件非常经济。用 NC 机床加工出的单件质量好且加工快。

　　5）缩短非加工时间。花在加工零件上的时间多了，用于运输和等待的时间短了。

　　6）降低刀具成本。数控机床在多数情况下无需复杂夹具。

　　7）降低检验与装配成本。产品质量提高了，因而减少了检验的必要，保证零件能按需装配。

　　8）增加机床使用时间。由于能迅速自动地更换工件与刀具，机床闲置时间缩短了。

　　9）便于加工复杂形状的工件。新型控制器功能多、编程能力强，因而轮廓加工与复杂形状的加工非常方便。

　　10）减少零件存货。人们能根据穿孔带上的信息加工所要求的零件。

　　自从约 200 年前的第一次工业革命以来，NC 已对工业界产生了深远的影响。计算机的发展与 NC 技术的应用拓展了人类的思维与体能。NC 装置将符号式输入转换为可用输出，将人们的想法转换为具有创造性、生产性的结果。NC 技术已经取得了迅速发展，它几乎应用于制造业的每一个领域，如机加工、焊接、冲压加工和装配。

　　如果工业规划合理、逻辑正确，与第一次工业革命一样，第二次工业革命将对人类社会产生同样甚至更大的影响。在我们从事 NC 不同阶段的工作时，必须牢记 NC 技术贯穿于整个加工过程的始终。

　　回顾过去的发展历史，可以肯定，未来的制造业属于计算机辅助制造（CAM）与计算机集成加工（CIM），自动化工厂在不远的未来将成为现实。

　　虽然起源于航空工业，但 NC 在制造业中被广泛接受。很明显，在大多数金属加工与制造业中，CNC 机床的应用持续增长，航空、国防、汽车、电子、仪器以及机床行业中都用了数控机床。微电子技术的发展降低了 CNC 设置的成本，在工具、模具与制模车间使用 CNC 机床已很常见。随着低成本 OEM（设备制造商）的出现与立式 CNC 铣床的改进,甚至那些专门生产某种标准工件的车间也会使用 CNC。

虽然数控机床传统上是指金属切削机床,但折弯机、成形机、冲压机和测量机也使用数字控制系统。

阅读材料 <center>**数 控 机 床**</center>

老式机床是按操作工站立在机床前进行操作来设计的。现在不需要这种设计了,因为在 NC 中操作工并不直接操控机床的运行。在传统机床加工时,只有约 20% 的时间用于切削材料。随着电子控制的加入,实际用于切削金属的时间已占到 80% 甚至更高,并且减少了将刀具定位到各加工位置所需的时间。

在过去,机床越简单越好,以便降低成本。由于不断上升的劳动力成本,人们开发出更好的机床,并配套电子控制设备,使该行业能够生产更多、更好且价格较低的产品,以和国际上的产品相竞争。

从最简单到最复杂的机床都会用到数控技术,最常见的机床有:单轴钻床、卧式车床、铣床、车削中心以及加工中心。

单轴钻床

单轴钻床是最简单的数控机床之一。多数数控钻床可在三个坐标轴上编程:

a. X 轴控制工作台左右运动。

b. Y 轴控制工作台靠近或离开立柱。

c. Z 轴控制主轴上下运动,确定孔的加工深度。

卧式车床

卧式车床是生产效率最高的机床之一,它是加工回转体零件时非常有效的工具(如图 2-1 所示)。大部分卧式车床可在两个坐标轴上编程:

a. X 轴控制刀具横向运动(切入或切出)。

b. Z 轴控制刀架靠近或离开主轴箱。

铣床

铣床一直是工业中最常用的工具之一(如图 2-2 所示)。像铣削、成型加工、齿加工、钻、镗、铰等只是可在铣床上进行的一小部分加工。铣床可在三个坐标轴上编程:

a. X 轴控制工作台的左右运动。

b. Y 轴控制工作台靠近或离开立柱。

c. Z 轴控制升降台或主轴垂直(上或下)运动。

车削中心

研究表明,整个金属材料的切削操作大约 40% 是车床上进行的。早在 20 世纪 60 年代中期,人们就开始研制车削中心。这种数控机床具有比普通车床更高的加工精度和效率。一般数控车削中心仅在两个坐标轴上工作:

a. X 轴控制转塔头横向运动。

b. Z 轴控制转塔头纵向运动(靠近或离开主轴箱)。

加工中心

加工中心是20世纪60年代发展起来的。有了加工中心，人们不必把零件从一台机床转移到另一台机床就能完成各种加工。由于工件经过一次装夹后便能进行多种加工，所以大大提高了生产效率。加工中心主要有两类：卧式加工中心与立式加工中心。

a. 卧式加工中心可在三个坐标轴上工作：
（a）X轴控制工作台左右运动。
（b）Y轴控制主轴垂直（上或下）运动。
（c）Z轴控制主轴水平运动（切入或切出）。

b. 立式加工中心在三个坐标轴上工作：
（a）X轴控制工作台左右运动。
（b）Y轴控制工作台靠近或离开立柱。
（c）Z轴控制主轴垂直（上或下）运动。

第二部分　数控机床刀具

第3单元

课文　　　　　　　　　　刀 具 系 统

　　加工中心是一种多功能机床，可用钻头、丝锥、铰刀、面铣刀、镗刀等各种刀具对工件进行多种加工。为了使这些刀具能迅速、准确地插入机床，所有刀具必须有相同的锥柄与机床主轴适配。数控加工中心主轴上最常用的锥柄是 No.50，这种锥柄具有自释放功能。刀柄还必须有凸缘或卡圈，以便换刀机械手夹持；还应有专打拉钉、螺孔或其他设备，以通过电动拉杆或其他夹紧机构将刀具牢固地夹紧在主轴上。

　　准备加工工序时，要用刀具装备图选择加工零件所需的全部刀具，然后将每把刀具预先装备到合适的刀柄，并预设正确长度。一旦所有刀具装设就绪，便可装入机床刀库的指定刀位，以便根据零件程序要求自动选刀。

刀具识别

数控机床使用多种方法识别加工用的各种刀具，最常用的刀具识别方法有：

1. 刀具刀位法

早期加工中心在刀库中给每把刀具分配一个指定的刀位，零件程序可以调用任何刀具。

2. 刀柄编码环方法

由特定的交互式识读机根据刀柄上特定的编码环来识别刀具。

3. 刀具装配号方法

多数现代机床控制单元具有一种刀具识别特征，即允许零件程序使用5~8位刀具装备号码从刀库中调用刀具。

每一个刀具装备号码都与刀库中的一个特定存储位置对应，其位置可由穿孔带设定，也可由操作人员用机床控制单元设定，或通过刀具管理遥控平台设定。

刀具管理程序

为了达到最大生产能力，给机床配置功能齐全的刀具管理程序就显得十分重要。只有每步操作都能用上最合适的刀具，性能最优越的数控机床才有可能接近其潜在的加工能力。刀具管理程序应涵盖多种信息：刀具设计类型、标准代码体系、外购状况、行之有效的加工技巧、有效的零件程序、刀具在机床上的最佳应用等。

良好的刀具原则应包括以下内容：

1. 标准规则

必须建立刀具的标准规则。

从刀具工程师、零件编程员、高级管理人员到安装工、机床操作工，人人都必须弄清楚这个规则。

必须明确每个人在选刀时担当的角色。

2. 刀具尺寸标准

所有外购、特制刀具必须符合公认的刀具尺寸标准。

必须重磨刀具时，应将其修磨到下一级 NC 尺寸标准。

零件编程员必须按刀具标准准则进行编程。

3. 刚性切削刀具

尽可能选用最短的刀具完成每项工作，以确保定位精度与刚度。

刀柄应为一体化结构，以提高刚度。

4. 刀具准备

在刀具安装、刀具补偿、重磨刀具时，相关人员必须严格遵守有关规定。

必须明确每项工作中各人的职责，确保没有争议或误解。

5. 可转位刀片刀具

尽可能使用硬质合金刀具，这种刀具抗磨损性好，生产效率高，尺寸精度高。

立方氮化硼（CBN）刀片用于硬质合金刀具无法满足加工要求的硬铁合金加工。

人造金刚石（多晶）刀片用于加工非铁材料。

刀具管理程序的好坏在很大程度上取决于零件编程员。为使工作更有效，编程员必须全面掌握实际加工知识、加工流程及每道工序所选刀具类型。为使刀具管理规则能更有效地发挥作用，大多数现代 CNC 装置都带有标准功能、选配功能及相关程序。

阅读材料　　　　　　　　　**切　削　刀　具**

在加工中心上精确加工零件，基本的要求是为每一步加工操作选择合适的刀具。但就每个具体工件的加工而言，选择刀具时一般没有完全的准则可供遵循。因此，只有 NC 编程员熟练掌握切削刀具及其应用知识，方可对任何零件正确编程。

加工中心使用多种刀具完成各种加工操作。按材料分，这些刀具有常规高速钢刀具、硬质合金嵌入式刀具、CBN（立方氮化硼）复合刀片嵌入式刀具、多晶金刚石嵌入式刀具。按功能分，常用刀具有面铣刀、钻头、丝锥、铰刀、镗刀等、研究表明，在加工中心上，平均每个加工周期的工作时间中有20%进行铣削、10%用于镗孔、70%进行孔加工。普通铣床的切削时间大约占总加工时间的20%，而加工中心的切削时间可高达75%。结果刀具利用率高了、寿命缩短了，导致刀具的消耗量增大。

面铣刀

面铣刀与空心面铣刀是加工中心上广泛使用的两种刀具，它们能完成多种加工操作，如铣削表面、铣槽与铣轮廓，刮孔口平面，反镗孔，用圆弧插补对孔进行粗加工与精加工。

钻头

通用钻头与专用钻头均可用于孔加工。随着钻头直径与长度的增加，孔径误差与孔位误差也随之增加，因而孔加工时总是选用能加工出所需深度的最短钻头。若在加工中心上钻孔，推荐使用粗短钻。

中心钻

在孔加工中，可以先用中心钻预加工出孔的准确位置，然后再进行其他加工。但是使用中心钻进行孔加工有其缺陷，即若使用时不注意，细小的导向钻头容易崩断。这时可用定位工具代替中心钻，它有一个90°内角，广泛应用于孔定位中。

丝锥

在设计丝锥时，应使丝锥能够承受攻螺纹所需转矩，并能清除孔内的切屑。攻螺纹是最难进行的加工之一，主要原因如下：

1）排屑不彻底；
2）切削液供给不足；
3）不同材料，螺纹牙型的粗、细不同；
4）攻螺纹的速度及进给量受螺纹导程影响；
5）所需攻螺纹的深度。

铰刀

铰刀有多种型号与尺寸。铰刀是一种端部可旋转的切削刀具，主要对预先钻好或镗好的孔加工到准确尺寸，并达到良好的表面粗糙度。

镗刀

镗孔是把预先钻过、镗过或用中心钻加工过的孔扩至准确尺寸与位置并达到所需表面粗糙度的一种操作。这项加工通常用单刃镗刀完成。在选取镗杆时，必须审慎考虑其长度与直径，因为随着长度与直径比的增加，镗杆刚度会有所降低。例如，一个长径比为1∶1的镗杆，其刚度比长径比为4∶1的镗杆刚度高64倍。

第4单元

课文　　　　　　　　　　　**刀具半径补偿**

什么是刀具半径补偿

刀具半径补偿是一种调节刀具半径的行为，其目的是使加工出来的工件具有正确的几何形状。

刀具半径补偿是如何工作的

了解 CNC 控制器在施加刀具半径补偿时如何编译运动，将十分有助于了解如何在程序中使用半径补偿并避免出错。简单地说，程序员需要告诉控制器（通过 G 命令）加工过程中铣刀和工件的位置关系。加工轮廓时，刀具或者在工件左边（G41），或者在右边（G42）。合理设定刀具半径补偿后，铣刀在完成程序中要求的所有动作时，控制器都要求它处于刀具的左边或者右边（根据程序员的选择）。这可能有点难以想象，特别是对初学编程者而言。CNC 控制器会自动让刀心偏离程序轨迹，偏移的尺寸为特定刀具半径补偿偏置量。正因为如此，我们才可以直接对工件表面轮廓进行编程，而将计算刀心轨迹的工作留给控制器。

在本例中，在从点 1 到点 2 的过程中设置了半径补偿（注意点 1 是刀心位置）。设置半径补偿后，注意控制器是如何自动使刀具偏移到运动路线的右边的（如图 4-1 所示）。控制器应该让刀具处于工件表面轮廓右侧多远的地方，则由刀具半径补偿偏置（刀具半径）确定。

对初学者而言，有两个问题不好理解，关于刀具半径补偿的所有问题几乎都源自这两个问题。首先是设置移动（从点 1 到点 2）。设置补偿之前，必须将刀具置于一个位置，这个位置与第一个加工面间的距离应大于刀具半径（在本例的点 1 处）。其次是刀具接近工件和回退到安全位置的运动。注意，点 2 到点 3 的前一部分移动和点 13 到点 14 的移动实际并不是工件加工所需要的，但却必须包括在工件表面运动轨迹之内。

有了半径补偿，轮廓的粗加工和精加工就可以用相同的工件表面进行编程。但是，刀具半径补偿只能用于铣削中。尽管具有局限性，但是它不需要程序员计算两套坐标，正是这一原因使得半径补偿被大多数手工编程员所采用。

阅读材料　　　　　　　　　**刀具长度偏置及零点预置**

程序中用不同的刀具进行加工时刀具的长度是不同的。工业中对于修正刀具长度的不同通常有两种不同的方法，即使用偏置和使用零点的预置。

刀具长度偏置

刀具长度偏置用于单一程序进行换刀时，它可用来修正不同刀具的长度及位置，这样操作所有刀具的时候就可以认为它们的长度和位置都是相同的。根据刀具代码，用于修正的偏置量从控制器的刀偏寄存器中调出来。

如图 4-2 所示，这是一个具有 3 个钻头的车床转塔刀架。在理想情况下，每当一个刀具接触到工件时，Z 轴寄存器就会读入一个零值。因为每把刀具都有不同的长度，所有 Z 向偏置量都要做正向或负向调整。这样，编程的时候就可以认为它们的长度都和 1 号刀具一样。

在图 4-2 中，1 号刀具是标准的，而 2 号和 3 号刀具是要与它进行比较的。沿着 Z 轴方向，2 号刀具比 1 号刀具长 0.50。因此，存储在 Z 轴偏置寄存器的偏移量为 -0.50。

当刀具代码 T0202 和换刀指令 M06 一起被读取后，机床将会自动将刀具移动到补偿位置——即将 2 号刀具放在 1 号刀具未经补偿的位置上。偏置量将一直得到应用直到被一个新代码所取消。

3 号刀具比 1 号刀具短，因此它的 Z 轴偏置量相对于 1 号刀具来说就是 +1.34。

零点预置

一般来说，有一种方法能将车床的程序参考零点设定在（刀具）出发点，而并不需要真的重新设置机床轴的寄存器。通过一个代码或字，控制器能将刀具目前所处的位置设定为零点，这样就可以完成上面所说的工作。G50 代码后面跟一个距离就可以建立一个程序参考零点，该距离是指程序参考零点到刀具当前位置之间的距离。控制器读入调整了新的程序参考零点的 G50 代码，刀具当前的位置作为暂时的参考点。

例如：

G50 X10 Z60

这个命令通知控制器刀具目前距离程序参考零点的位置是 X 方向为 10 毫米，Z 方向为 60 毫米。于是控制器就知道这个操作的程序参考零点位于 Z 向距离卡盘 60 毫米处，X 向位于中心。

第三部分　数控机床结构

第 5 单元

课文　　　　　　　　　　　**MCU 与 CPU**

MCU 是整个数控机床运行的中枢环节，其主要任务是接收零件程序并将这些信息转换为机床可以识别的语言，从而实现加工成品所需要的各种功能。这就要求能用电气与液压伺服设备正确控制继电器或螺线管的开关状态及机床运行过程。

数据译码与控制

从穿孔带获得二进制编码数据（BCD）并转换为二进制数，这是 MCU 的首要操作之一。下一步是把这些信息送到 MCU 的保存区，通常称为缓冲存储器（缓冲区）。缓冲区的作用是把这些信息或数据迅速传到 MCU 的其他部分。如果没有缓冲存储器，MCU 只有等待读带机完成译码并发出下一组指令。这就会暂时中断信息传输，使机床暂停运行，从而在工件上留下刀痕。没有缓冲存储器的 MCU 必须配有高速读带机，以免信息传输与机床运行中出现中断现象。

MCU 的数据译码与控制设备可处理所有由穿孔带设定的机床运动的信息。该设备也允许操作者通过控制面板手动修改程序。

MCU 的发展

自从 20 世纪 50 年代初期以来，MCU 已从笨重的真空管单元发展成为今天的计算机控制单元，其中已经融进了最新的微处理技术。20 世纪 70 年代初期以前，MCU 的各项功能，如纸带格式识别、绝对定位与增量定位、插补、代码识别等，均由 MCU 的电子元器件实现。这种 MCU 曾被称为硬控制器，因为这些功能已经固化在 MCU 的计算机部件中而无法改变。

20 世纪 70 年代中期，软控制器的发展使 MCU 更具有柔性且成本更低。简易型计算机部件，甚至小型计算机成为 MCU 的组成部分。过去曾由制造商固化在硬连线系统中的那些功能，现在都已包含在 MCU 软件中。这种计算机的逻辑能力更强、价格更低，并能对多种功能进行编程。

CPU

任何计算机的 CPU 均由三部分组成：控制器、算术—逻辑单元及存储器。CUP 的控制器是计算机的最主要组成本分。CPU 各组成部分的主要功能如下：

1. 控制器
1）协调、管理整个计算机系统。
2）从存储器中获取程序数据并完成译码。
3）向 NC 系统其他单元发送信号，执行特定操作。
2. 算术—逻辑单元
1）完成程序要求的加、减、乘、计数等运算。
2）提供逻辑问题的答案（对照、决定等）。
3. 存储器
1）短期或临时存储正在处理的数据。
2）提高计算机主存储器中信息的传输速度。
3）有一个存储器寄存器，提供一个特定位置来存储或调用一个字。

阅读材料　　　　　　　　**机床运动及控制**

定位控制

数控加工的基础就是对机床的滑板的运动进行编程，使其达到预先确定的位置。下面描述了 3 种定位控制方式。

1. 点位定位

点位定位是指编写的只识别所需的下一个位置的指令。机床可通过一个或多个轴运动来达到所需位置。

2. 直线运动

直线运动控制也叫做直线插补。被编程的运动由相关的指令产生，这些指令确定了所需要的下一个位置信息，以及在到达那个位置的过程中要使用的进给速度。直线插补用线段把许多相邻或离得很远的被编程点连起来。

3. 轮廓定位

轮廓定位是指控制两个轴或多轴联动以保持恒定的刀具—工件相对位置的能力。NC 纸带上的编程信息能够使刀具从一点到另一点准确定位，并按程序设定的进给速度沿预定路径运动，从而加工出所要求的形状或轮廓。

刀具运动控制的回路系统

控制回路系统发出电信号来驱动电动机控制器并从电动机控制器中接收某种形式的电反馈。现在的 CNC 机床的运动控制系统主要有两种：开环系统和闭环系统。

1. 开环系统

开环系统（如图 5-1 所示）利用步进电动机来驱动机床运动。对于每一次接收的脉冲，这些电动机都会转过一个固定的角度，通常为 1.8°。步进电动机由 MCU 产生的电信号来驱动。它们与机床工作台上的滚珠丝杠和主轴相连。每接收到一个信号它们就会驱动工作台和（或）主轴移动一个固定的量。电动机控制器反馈信号表示电动机已经完成了指定的运动。但是，该反馈并不用于检测机床的实际运动与程序要求的精确运动的接近程度。

2. 闭环系统

在闭环系统中通常是用特殊的电动机，即所谓的伺服电动机来执行机床运动（如图 5-2 所示）。电动机的类型有 AC、DC 和液压型伺服电动机 3 种。作为功率最高的伺服电动机，液压型电动机常在大型 CNC 机床中使用。AC 伺服电动机的功率仅次于液压伺服电动机，通常应用在加工中心上。

伺服电动机不像步进电动机那样根据脉冲数运动。AC 和 DC 伺服电动机的速度是可变的，其大小取决于电流强度。液压型伺服电动机的速度则取决于流过它的液体量。来自 MCU 的电流强度决定伺服电动机的转速。

伺服电动机是和主轴相连的，它们同样通过滚珠丝杠与机床工作台相连。旋转变压器持续监控工作台和（或）主轴的移动量，并将该信息传回 MCU，于是 MCU 就可以调整信号使工作台和（或）主轴的实际位置不断接近程序规定的位置。这种能够反馈信号的系统就称为伺服系统或伺服机构。即便驱动电动机的功率范围很大，它们也能对刀具进行高精度定位。

近年来，开环系统在 CNC 中的应用重新引起了大家的兴趣。由于步进电动机的精确度和功率的提高，某些情况下就没有必要采用昂贵的反馈系统硬件及其关联线路。这种新系统可以大大地降低机床价格及其维护成本。

第 6 单元

课文　　　　　　　　　　加工中心的种类与组成

加工中心分为两大类：卧式主轴加工中心和立式主轴加工中心。

1. 卧式主轴加工中心

1) 移动立柱加工中心通常配有一个或两个工作台来装夹工件。使用这种加工中心，操作员可在加工工件的同时，在另一个工作台装夹其他工件。

2) 固定立柱加工中心上配有一个平板架，这是一个加工工件的可移动工作台。完成加工后，工件与平板架一起移到滑台，然后滑台转动，把新的平板架与工件输送到加工位置。

2. 立式主轴加工中心

立式主轴加工中心呈鞍形结构,其中装有滑动导轨,该导轨采用立式滑块运动,而不是主轴运动。

CNC 加工中心的组成部件

CNC 加工中心的主要部件有床身、床鞍、立柱、工作台、伺服电动机、滚珠丝杠、主轴、刀库以及机床控制单元（MCU）。

床身——床身通常由优质铸铁制成,使机床具有刚性加工能力,可实现重型加工,并能保持良好的加工精度。在床身上安装的经淬火和磨削的导轨为各直线轴提供刚性支撑。

床鞍——床鞍装在经淬火和磨削的导轨上,为加工中心提供 X 轴方向的直线运动。

立柱——立柱装在床鞍上,具有很高的扭曲强度,以防止加工过程中的变形与偏移。立柱为加工中心提供 Y 轴方向的直线运动。

工作台——工作台装于床身上,为加工中心提供 Z 轴方向的直线运动。

伺服系统——伺服系统由伺服驱动电动机、滚珠丝杠与位置反馈编码器组成,它为 XYZ 轴滑台提供快速准确运动与定位。反馈编码器安装在滚珠丝杠端部,构成闭环系统,单向重复定位精度可达 ±0.001 英寸（0.025 4 毫米）。

主轴——主轴可以 1 转每分钟的增量编程,其转速在 20~6 000 转每分钟之间。主轴可以是固定式（卧式）,或是倾斜/轮廓主轴,以提供一个附加轴（A 轴）。

换刀机构——换刀机构有两种基本类型,即立式换刀机构与卧式换刀机构。换刀机构中存放了许多预选刀具,由零件程序自动调用。换刀机构一般是双向的,允许以最短路径随机存取刀具。实际换刀时间通常只有 3~5 秒。

机床控制单元——机床控制单元允许操作员进行各种操作,如编程、加工、诊断、刀具与机床监控等。根据不同的生产规格,控制单元也不相同。新型控制单元采用尖端技术,使机床性能更可靠,并使整个加工过程对人为技能的依赖越来越少。

阅读材料 **机床轴系统**

什么是机床轴？

axes 是 axis 的复数形式。它们都是运动和位置的参考基准。一根轴就是指一条中心线,CNC 的所有运动都是相对于该轴的,或者是以该轴为基准的。机床轴系统是机床运动的国际标准。

运动及方向

轴是用来识别运动的。所有的 CNC 机床都需要一些方法来识别程序中需要哪些运动。举例来说：对于立式铣床,程序员如果要使刀具左右运动,就要调用 X 轴;同样地,若需要工作台内外运动,就要调用 Y 轴。

方向通过符号值来确定。一旦坐标确定,它的方向就有一个正或负,从而决定了机床的运动方向为左或右、内或外、上或下、顺时针或逆时针。

以图 6-1 所示的铣床为例,铣床工作台上 X 轴的方向代表左右运动方向,编程时如果要使刀具向右运动,则需要发出"X 轴正向移动"的命令。同理,向左运动需要"X

轴负向"；而刀具向内（远离操作员）时为 Y 轴正向，刀具向上为 Z 轴正向。

坐标系

注意图 6-1 中铣床的三根轴相互之间成 90°夹角，这就是所谓的直角坐标系。直角就是成 90°夹角。绝大多数 CNC 机床的标准坐标系都是成直角关系的。

右手法则

绝大多数的 CNC 机床都可以应用图 6-2 所示的右手法则来确定坐标系。当右手的拇指指向 X 轴正向时，食指方向即为 Y 轴正向，同时 Z 轴正向由中指确定。

机床轴运动分为两种单一的类型，要么是直线（线性运动），要么是圆周（旋转运动）。每种运动都要以轴为基准。

轴是一条可以作为运动和距离的基准直线。旋转运动时（机床）是以对称轴为中心的，这一点与车轮绕车轴运动相似；直线运动与基准线是平行的。

第四部分 数控机床系统

第 7 单元

课文　　　　　　**FANUC 系统操作单元——CRT/MDI 面板**

图 7-1 所示为 MDI 面板的标准键盘。

MDI 面板各按键的功能如下：

1. 电源开/关键

该键用于打开或关闭 CNC 电源。

2. 复位键

按下此键可以复位 CNC 系统、取消报警等。

3. 地址和数字键

按下此键可以输入字母、数字和其他字符。

4. 功能菜单键

位置显示键：在 CRT 显示屏上显示机床现在的位置。

程序键：在编辑模式下，编辑并显示内存中的程序。在手动数据输入（MDI）模式下，输入并显示 MDI 数据。

偏置功能键：刀具偏置模式选择，用于刀具偏置设置。

菜单键：显示操作菜单。

诊断/参数键：设定和显示自诊断表和参数的内容。

操作/报警键：显示报警号。

辅助/图形键：进行图形模拟。

宏指令键：设置宏指令功能。

5. 换档键

一些地址键上有两个字符，按换挡键可以转换字符。按下换挡键后，"-"显示在屏幕光标处。这表明通过按下地址键，按键右下角的字符能被输入。

6. 光标移动键

下面描述了两种类型的光标移动键。

→：该键用于向前移动一小段光标。

←：该键用于向后移动一小段光标。

7. 页面转换键

下面描述了两种页面转换键。

↓：该键用于在 CRT 显示屏上向前转换页面。

↑：该键用于在 CRT 显示屏上向后转换页面。

8. 取消键

按下该键删除输入到按键输入缓冲器中的最后一个字符或符号。按键输入缓冲器的按键内容显示在 CRT 显示屏上。当地址键或数字键再一次按下后，字母或数字插入的位置显示"-"。当按下取消键（CAN），在"-"前面的字符立即被删除。

9. 程序编辑键

修改键：按下这个键修改上一个输入到按键输入缓冲器中的字符或符号。

插入键：在当前光标前面插入字符或符号。

删除键：删除指定的字符、符号或程序。

10. 输入键

当地址或数字键按下以后，字母或数字输入到按键输入缓冲器中，并且显示在 CRT 显示屏上。利用输入键在偏置寄存器中设置输入到键盘输入缓冲器中的数据。

11. 输出启动键

按下该键后，CNC 开始向外部设备输出内存中的程序或参数。

阅读材料　　　　　　　**FANUC 系统操作面板**

图 7-2 所示为 FANUC 系统操作面板的正面外观，各键含义见表 7-1。

表 7-1　各键含义

键	解　释
AUTO	自动模式选择：自动加工模式。
MDI	手动数据输入模式选择：手动数据输入模式。
EDIT	编辑模式选择：进行程序编辑。
JOG	手动进给模式选择：手动进给模式。
IN JOG	点动进给模式选择：点动进给。
MPG	手轮进给模式选择：手轮进给模式。
HOME	回零模式选择：各轴回零。
TEACH	手动（手轮示教）模式选择：手动示教（手轮示教）模式。

(续)

键	解　释
OFSET MESUR	刀具偏置模式选择：设置刀具偏置。
？NC	CNC警报：当出现CNC警报，发光二极管亮。键盘没有任何含义。
？MC	机床警报：当出现机床警报，发光二极管亮。键盘没有任何含义。
SINGL BLOCK	单段模式：单段运行模式程序，进行程序测试。
PRG STOP	程序停止（仅用于输出）：执行到M00命令，程序停止运行，发光二极管亮。
OPT STOP	任选停止：按下此键，碰到M01命令，自动停止运行。
DRY RUN	空运行：信号接通后，轴处于点动进给状态，不进行工件的切削，检验刀具的运动。
PRG TEST	机床锁定：信号接通后，CRT屏幕显示自动运行轨迹，机床不运动。用于检验程序。
MPG X	手轮进给选择X轴：手轮进给模式下，接通这个按钮。用于检验程序（Y轴、Z轴、C轴和第四轴与X轴相同）。
WORK LIGHT	工作灯：工作灯开关控制。
MPG INTRT	手轮中断：自动运行中选定此键，手轮进给增加到程序中。
AXIS INHBT	轴的停止：指定轴或所有的轴停止运动。
LOW ×1	手动进给修调开关，向上：对手动进给的五个级别进行修调。
MEDL ×10	手轮进给倍率：手轮进给倍率依次为×1、×10、×100、×1000。
+X	手动进给方向：在手动进给（或点动进给）时，按下这个按钮，在手动进给（或停止进给）选择移动轴移动的方向（−X，+Y，−Y和+Z与+X相同）。
TRVRS	横向进给：点动进给中按此按钮，实现快速横向进给。
CYCLE START	循环开始：自动运行开始。
CYCLE STOP	循环停止：自动运行停止。
CLNT ON	切削液开：开始加入切削液。
CLNT OFF	切削液关：停止加入切削液。
CLNT AUTO	自动冷却：切削液的开关自动控制。
SPDL 100%	主轴倍率100%：主轴转速倍率100%。
SPDL INC	主轴转速增加：主轴电动机加速运转。
SPDL DEC	主轴转速降低：主轴电动机减速运行。
SPDL CW	主轴正转：主轴顺时针旋转。
SPDL CCW	主轴反转：主轴逆时针旋转。
SPDL STOP	主轴停止：主轴旋转停止。
SPDL JOG	主轴手动进给：主轴进给设置为手动。

第8单元

伺服控制

课文

　　伺服控制由电力、液压与气动装置组合而成，其作用是控制机床工作台的运动。最常用的伺服控制系统有开环系统与闭环系统两种。

在开环系统中，纸带阅读机译出穿孔带上的信息，并将其暂存起来，直到被机床使用。然后，纸带阅读机将这些信息转换成电脉冲信号并送至控制单元，以驱动伺服控制单元；伺服控制单元按照纸带提供的信息驱动伺服电动机，实现特定功能。各伺服电动机的转动量取决于伺服控制单元发出电脉冲的数目。NC 机床通常采用 10 头螺纹/英寸的精密丝杠。如果与丝杠相连的伺服电动机收到 1 000 个电脉冲，则工作台移动 1 英寸（25.4 毫米），因而一个脉冲使工作滑台移动 0.001 英寸（0.025 4 毫米）。开环控制系统相当简单，但由于无法检测伺服电动机是否正确执行其功能，因而在精度要求高于 0.001 英寸（0.025 4 毫米）的场合，通常并不被使用。

开环系统好比一位射手，虽然他尽其所能计算如何命中远程目标，但却没有一位观察者来确认射击是否准确。

闭环系统也可以比作同一位射手，但现在有一个观察者来确认射击的准确性。他将有关射击准确性的信息传递给射手，射手再做必要的调整以便击中目标。

除了在电路中加入反馈装置之外，闭环系统与开环系统在其他方面很相似。反馈装置通常叫做传感器，能将伺服电动机驱动工作台移动的量与控制单元发出的信号进行比较。控制单元使伺服电动机不断做出必要调整，直到来自控制单元的信号与伺服装置的信号相等为止。闭环系统中，10 000 个电脉冲才能使工作台移动 1 英寸（25.4 毫米），因而该系统的一个脉冲使工作台仅移动 0.000 1 英寸（0.002 54 毫米）。闭环数控系统之所以控制精度很高，就在于它能记录指令信号，并能自动补偿误差。如因切削力而使工作台位置发生偏移，反馈装置会显示这一偏移，机床控制单元（MCU）会自动进行必要调整，从而使机床工作台返回原位。

阅读材料　　　　　　　　**FANUC-BESK 数控系统**

FANUC-BESK 系统 3M-A 型用于两轴或三轴轮廓控制的数控钻床和数控铣床。实际上，3M-A 型系统是经济型数控系统。因此，它最适合传统的数控铣床。

在控制电路中，由于使用高速微处理器和大量的特别传感器，因此数字电路元器件锐减，而数字电路板也减少到只剩一块。其结果是使得数控机床具有显著的高可靠性。与此同时，由于控制部分所需电源非常小，所以可以将其很方便地合并进一个电磁控制箱，放于机床的一侧。

FANUC-BESK 系统 3M-A 型是面板装配结构，因此，它很方便地实现了机电一体化。此外，可编程序控制的使用简化了电磁控制电路，使得数控系统更简单、更集体化。

FANUC-BESK 直流伺服电动机广泛应用于世界范围的数控系统中，可以实现高速、强效、稳定的加工。此外，FANUC-BESK 直流主轴电动机可以实现电主轴的定位控制，因此比以往更加有效。

手动脉冲发生器的轴控制与回放

通过三个手动脉冲发生器，机床就能实现两轴联动的手动操作，像普通铣床一样。回放功能可以记录下手动试切削路径，以便自动加工中使用。

手动脉冲发生器可以增加到100个，所以操作者可以像操作普通铣床一样，手动操作数控机床。

高性能，高可靠性

在控制电路中广泛使用高速微处理器和大量的特制传感器，数字电路元器件数量锐减，而且控制部分只使用了一块数字电路板。

高效电源的使用减少了机床电源发电机。数控操作面板的键盘转换开关使用了专门的橡胶盖，以防止灰尘。即使伺服系统使得机床产生位置偏移，机床也能进行自动补偿达到正确的运动位置。此外，出厂前仔细的结构检查和全面的性能测试确保了机床长期无故障运行。

方便维修

FANUC-BESK 系列 3M-A 型的维修非常容易，原因如下：

1）微处理器始终监测 NC 内部的运行状态，并且将其分类显示出来。NC 发生故障时，警报灯亮，NC 停止运行，详细的警报信息被分类显示出来。

2）所有输入、输出 NC 的开/关信号都有所显示。

3）任何一种输出 NC 的开/关信号都可以通过点动式方式下的手动数据输入进行手动操作。

4）各种参数设备，如加/减速的时间常数、横向快速进给常数，都能被显示出来。

第五部分　CNC 编 程

第 9 单元

课文　　　　　　　　　　　程序设计概念

在完全理解 CNC 之前，你必须首先明白制造企业是如何在 CNC 机床上完成加工工作的。下面的例子说明企业是如何细化其 CNC 过程。

CNC 工作流程

1. 获取或开发零件图。
2. 选择加工零件的机床。
3. 制定加工工序。
4. 选择所需的加工刀具。
5. 计算编程坐标。
6. 根据加工刀具和零件材料计算转速和进给量。
7. 编写数控加工程序。
8. 准备步骤清单和刀具清单。
9. 将程序输入机床。
10. 修改程序。

11. 若无修改，运行程序。

准备程序

CNC 机床执行的程序是连续的加工指令。这些加工指令是一些 CNC 代码，包含了加工零件的全部信息，由程序员编写。

CNC 代码包括许多程序段（也可以称为程序行），每一段都包含了一种运动或一个准确动作的单个命令。像传统机床一样，一个动作接一个动作的进行，因此 CNC 代码以数字段的形式被罗列出来。

下面是数控铣床的程序实例。注意每一段程序是怎样编写的，而且通常每一程序段中只有一个专用指令。也要注意程序段的编号是以 5 为增量（这是编程软件启动时的默认设置）。每个程序段都包含专用信息，指导机床按顺序加工。

工件尺寸：X4，Y3，Z1

刀具：3 号刀，3/8 英寸钻头

刀具起始点：X0，Y0，Z1.0

%	（程序开始标记）
:1002	（程序号 1002）
N5 G90 G20 G40 G17	（第五程序段，绝对坐标、英制编程）
N10 M06 T3	（更换 3 号刀）
N15 M03 S1250	（主轴正转，转速 1 250 转每分钟）
N20 G00 X1.0 Y1.0	（快速进给到 X1.0，Y1.0）
N25 Z0.1	（快速退刀到 Z0.1）
N30 G01 Z-0.125 F5	（切削进给到 Z-0.125，进给速度 5 英寸每分钟）
N35 X3.0 Y2.0 F10.0	（直线进给到 X3.0，Y2.0，进给速度 10 英寸每分钟）
N40 G00 Z1.0	（快速退刀到 Z1.0）
N45 X0 Y0	（快速移动到 X0，Y0）
N50 M05	（主轴停转）
N55 M30	（程序结束）

CNC 代码

每一个程序都包括两种主要的 CNC 代码类型，或者称为字母地址。主要的 CNC 代码有 G 代码和 M 代码。

G 代码是准备功能，包含实际的刀具运动（如机床的控制）。这些运动包括快速运动、进给运动、圆弧进给运动、暂停、粗加工循环和精加工循环。

M 代码是辅助功能，包括加工必需的，但不是实际的刀具运动（如辅助功能）。这些动动包括主轴的启动与停止、刀具的更换、切削液开与关、程序停止以及相关功能。

其他的字母地址是 G、M 代码中用于生成指令的变量。大部分的 G 代码包含一个变量，由程序员定义，实现特定的功能。在 CNC 程序中每一个指令都称为一个字母地址。

程序中用到的字母如下：
N　程序的段序号：指定一程序段的开始
G　准备功能，如前面说明
X　X轴坐标
Y　Y轴坐标
Z　Z轴坐标
I　圆心的X方向坐标
J　圆心的Y方向坐标
K　圆心的Z方向坐标
S　设置主轴转速
F　设置进给速度
T　使用指定刀具
M　辅助功能，如前面说明
U　X方向增量坐标
V　Y方向增量坐标
W　Z方向增量坐标

阅读材料　　编程基础

孔加工编程

与孔加工有关的程序编制是编程中最简单的，其中包括钻孔、镗孔、攻螺纹和锪孔。编程之所以简单，是因为程序员只需确定孔中心的坐标以及机床在该处的运动类型即可。如果使用得当，固定循环可以接管并指导机床完成所需运动。控制器中储存了许多固定循环，需要时，这些固定循环可以被程序调用，这样能节省编程的时间及所需磁带的长度。

固定循环是由一个程序段内的下列信息构成的：X和Y坐标，Z向参考平面和Z向最终深度。为了能更容易地理解固定循环，请参看如图9-1所示的G81循环。

G81循环命令机床：
（1）把刀具从Z的初始位置快速移动到参考平面R。
（2）以F为进给率，钻孔深度Z。
（3）快速返回到Z轴初始位置或参考平面R。
（4）如果在下一程序段中给出下一中心孔的X、Y坐标，则快速移动到该孔中心。

直线轮廓编程

直线轮廓加工指加工的轮廓完全由直线组成，这些直线可以是水平的、垂直的或任意角度的。在这个过程中将会用到直线插补。在零件编程中直线插补用来使刀具产生沿起点到终点的直线运动。直线插补模式专门用于实际的材料切除，比如加工轮廓、凹槽、面铣削以及许多其他的加工运动。

在线性插补的模式下，可产生3种运动类型：

水平运动，单轴；

垂直运动，单轴；

角度运动，多轴。

直线轮廓铣削如图 9-2 所示。

G01：指定直线插补模式。刀具以程序要求的进给率沿直线运动。

Z_n：n 指定切削的绝对深度。

F_n：n 指定刀具进入材料的进给速度，程序中后续的直线运动均采用此速度。如果没有输入，系统将使用程序给定的最新进给速度。

X_1Y_1：指定刀心在 1、2 等线的终点时的绝对坐标。

第 10 单元

课文　　　　　　　　　　　　G 代码

G02 圆弧插补（顺时针，如图 10-1 所示）

格式：　N　G02　X　Y　Z　I　J　K　F　（I，J，K 指定半径）

或：　　N　G02　X　Y　Z　R　F　　　（R 为指定半径）

圆弧插补俗称半径（圆弧）进给运动。G02 命令用于指定所有的顺时针圆弧插补运动，不管二次圆弧、部分圆弧还是整圆，只要它们在同一平面内，就可以用 G02 完成。G02 是 模态指令，并且由用户定义进给速度。

例如：G02　X2　Y1　I0　J-1

G02 命令需要有终点坐标和半径来切削圆弧。圆弧起点（X1，Y1），终点（X2，Y2）。为了找到半径，简单地测量一下起点和圆心的相对距离。半径用 X 和 Y 的距离表示。为了避免混淆，这两个值分别用变量 I，J 指定。

例如：G02　X2　Y1　R1

也可以通过输入圆弧的终点 X，Y 的坐标和半径 R 来指定 G02。确定半径值（I，J 的值）的比较简单的方法是列一个小图表：

中心点　X1　Y1

起点　　X1　Y2

半径　　I0　J-1

找到 I，J 的值比初看起来要容易一些。按照以下步骤进行：

1. 写出圆心点的 X，Y 坐标。

2. 在该坐标下，再写出圆弧起点的 X，Y 坐标

3. 画一条直线分离两个区域，将上面数值作减法。

结果：G02　X2　Y1　I0　J-1　F5

4. 计算 I 的数值，将圆弧的圆心和起点的 X 坐标相减。在本例中，X 坐标都是 1。因此，在它们之间没有差值，所以 I 为 0。计算 J 的值，因为 Y2 和 Y1 的差值为 1，方向向下，因此 J 为 −1。

G03 圆弧插补（逆时针，如图 10-2 所示）

格式：N　G03　X　Y　Z　I　J　K　F　(I, J, K 指定半径)

或：　　N　G03　X　Y　Z　R　F　　(R 为指定半径)

G03 命令用于制定所有逆时针圆弧插补运动，不管二次圆弧、部分圆弧还是整圆，只要它们在同一平面呢，就可以用 G03 完成。G03 是模态指令，并且由用户定义进给速度。

G04 暂停

格式：N　G04　P

G04 是非模态指令，它可以使所有的轴暂停一定时间，而主轴仍以指定的转速旋转。暂停主要用于钻削加工，目的是清除切屑。通常在进给切入和直线轮廓加工之间也使用暂停指令。这个指令需要指定持续的时间，用字母 P 表示，时间单位为秒。

阅读材料　　　　　　　　　　**M 代码**

M 代码是辅助功能，包括机床加工必要的但不是实际的刀具运动。也就是说，它们是辅助功能，如主轴的旋转和停止、刀具的更换、切削液的开关、程序的停止以及类似的相关功能。下面各部分描述了对应代码的具体功能。

　　M00　程序停止

　　M01　选择程序停止

　　M02　程序结束

　　M03　主轴顺时针转动

　　M04　主轴逆时针转动

　　M05　主轴停止

　　M06　更换刀具

　　M08　切削液开

　　M09　切削液关

　　M30　程序结束并返回起点

　　M98　调用子命令

　　M99　返回主命令

　　跳步运行　以"/"开头的程序段选择性地跳步运行

　　注释　注释可以以"（ ）"的 形式写在程序中

M00 程序停止

格式：N　M00

M00 指令是暂时程序停止功能。当该指令执行后，所有功能暂时停止，直到操作者重新输入指令。

在 CNCez 的模拟屏幕上会显示："程序停止，输入命令继续。"程序直到按下"输入"按钮才会继续运行。不同机床的提示符语句不同。

这个指令用在较长的程序中，以便停止程序进行清除切屑、测量尺寸、调整夹具或添加切削液等操作。

M02　程序结束

格式：N　M02

M02 指令表示主循环操作程序的结束。执行到 M02 指令，机床控制单元关闭所有机床操作（如主轴、切削液、进给轴、辅助运动）并结束程序。

这个指令写在程序的最后一行。

M03　主轴顺时针转动

格式：N　M03　S

M03 指令使主轴顺时针旋转。主轴的转速由 S 字母地址指定，单位是转数/分钟。

程序运行时，主轴的速度显示在程序状态窗口。主轴开/停状态显示在系统状态窗口（顺时针转动、逆时针转动或停止）。

M04　主轴逆时针转动

格式：N　M04　S

M04 指令使主轴逆时针旋转。主轴的转速由 S 字母地址指定，单位是转数/分钟。

程序运行时，主轴的速度显示在程序状态窗口。主轴开/停状态显示在系统状态窗口（顺时针转动、逆时针转动或停止）。

M05　主轴停止

格式：N　M05

M05 指令使主轴停止转动。尽管某些 M 代码可以关闭所有的功能（如 M00、M01），但是 M05 指令可以直接使主轴停止运动。M05 指令写在程序的结尾。

M30　程序结束，返回开头

格式：N　M30

M30 表示程序数据的结束。换句话说，其后再无其他程序指令。较早的数控机床由于不能区分前后程序，因此出现数据结束命令。M30 指令现在用于结束程序并返回程序的开头。

第六部分　计算机数控电火花加工

第 11 单元

课文　　　　　　　　　　**电火花加工**

利用电热能开发非传统（加工）技术，有助于提高切削性能差、难加工材料的加工经济性。通过控制一系列电火花的腐蚀来除去材料的工艺通常称为电火花加工，该工艺始于 1943 年的前苏联。随后的研究开发使电火花加工达到了现在的水准。

电火花加工的基本流程如图 11-1 所示。

当火花放电发生于阴阳电极之间时，热能集中于金属区并在放电区域使金属挥发。

为了提高效率，工件与刀具都浸在绝缘液体中（碳氢化合物或矿物油）。很明显，如果两个电极都是同一种材料，通常连接阳极的电极消耗的速率较快。因此，工件通常作阳极，刀具与工件的表面有一适当的距离，即是所谓的放电间隙。火花放电是以较高的频率、合适的电源而进行的。由于火花产生在刀具和工件的最接近点，且火花产生后最接近点发生变化（因为产生火花后材料被切除），所以电火花移动经过整个平面，最后，工件的表面（廓形）与刀具相吻合。这样，刀具在工件上产生所需要的形状。为了保证预设的放电间隙，通常会使用一伺服控制单元。火花隙能感知穿过它的平均电流并将该值与其现行值比较，其差值用于控制伺服电动机。有时，可用一梯级电动机代替伺服电动机。当然，很原始的操作也可能用一螺线管控制，如果这样机器就很便宜且构造简单。图 11-2 所示为一螺线管控制火花放电设备的排列。放电频率常在 200～500 000 赫兹范围，放电间隙常在 0.025～0.05 毫米间。火花隙间的峰值电压常在 30～250 伏之间，在加工过程中会产生高达每分钟 300 立方毫米的金属切除量，具体功率常在 10 瓦每立方毫米分钟左右。当使用一循环流动的绝缘液体时，可得到较高的效率和精度。最常用的绝缘液体是煤油，刀具常由黄铜或铜合金制成。

阅读材料 **激光束加工**

就像一束高速电子一样，一束激光同样可以产生很高的功率密度。激光是一道波长在 0.1～70 微米之间的电磁辐射很高的连贯束（在空间与时间上）。然而，机械加工的功率要求限制了它的有效使用波长范围在 0.4～0.6 微米之间。正因为激光束具有理想的相互平行与单色性，它能被聚焦于一很小的直径并能产生高达 107 瓦每平方毫米的功率密度。

为了产生高功率，常用脉冲红宝石激光器。连续 CO_2-N_2 激光发生器也成功地应用于机械加工中。

图 11-3 所示是一典型的脉冲红宝石激光器。一卷氙气闪光管环绕放置在红宝石标尺四周，容器周边的内置表面做成具有高度的反射性能，以使最大量的光落在红宝石标尺上而便于抽气操作。电容器充电并且一很高的电压施加在触发电极上以启动并发光。发射的激光束由一透镜系统聚焦，聚焦后的激光撞击在工件表面，通过汽化与高速的熔蚀切除一小部分材料。很小部分的熔融金属很快被汽化，产生坚固的机械脉冲，抛出大部分的液态金属。因为通过闪光管释放的能量比以激光束的形式放出的能量多得多，所以该系统应该被适当冷却。

激光束加工过程的效率很低，大约只有 0.3%～9.5%。典型的激光能量输出在 1 毫秒脉冲里是约 20 焦，最大功率值可达 20 000 瓦。光束的分离在 2×10^{-5} 戈左右，用一焦长 25 毫米的透镜，斑点直径达 50 微米左右。

像电子束一样，激光束也用来钻微孔或切割很窄的缝。激光可很容易地钻直径达 250 微米的孔，尺寸精度约为 0.025 毫米。当工件的厚度超过 0.25 毫米时，可得到每毫米达 0.05 毫米的锥度。

第 12 单元

课文　　　　　　　　**线切割电火花加工（1）**

电火花加工原理

电火花加工通常简称 EDM，它是采用电火花放电熔蚀工件来控制金属切削的加工方法。这样加工出的工件形状与切割刀具或电极的形状相反。加工时，电极与工件都浸入工作液中，常用的工作液是轻润滑油，此液体应是电的绝缘体。伺服机构保持电极与工件之间有大约 0.000 5～0.001 英寸（0.012～0.025 毫米）的间隙，以免二者互相接触。给电极加以频率约为 20 000 赫兹的低压大电流直流电，脉冲电能在电极与工件间的工作液中火花放电，从而在放电局部产生大量热量而使金属熔化，其微粒从工件上脱落下来。持续循环的工作液不仅冲走熔蚀后的金属微粒，并且有助于散除电火花产生的热量。

工业中使用的电火花加工机床分为两类：立式主轴电火花加工机床与线切割电火花加工机床（图 12-1a，b）。由于线切割电花火加工通常用数控编程来加工复杂形状，此处只详细介绍这类机床。

线切割电火花加工

线切割电火花加工是一种放电加工方式，加工时用数控运动在零件上产生所需轮廓或形状。它无需特制形状的电极，而是用连续运动且处于张紧状态的金属线作为电极。电极就是切割线，它可以用黄铜线、铜线或其他导电材料制成，其直径在 0.002～0.012 英寸（0.05～0.30 毫米）之间。切割线被控制在 X-Y 平面内沿工件轮廓曲线移动，从而在工件上电蚀出一条狭缝。切割线的这种控制运动是以 0.000 05 英寸（0.001 2 毫米）的微小增量连续进行的。这种方法可以非常准确地加工出任何轮廓形状，并可连续性重复加工许多零件。工作液通常是去除离子的水，它不断循环流动，冲走熔蚀的金属微粒。工作液不仅可以使切割线与工件之间保持合适的导电性，并且有助于减少火花产生的热量。

线切割电火花加工机床的组成部件

线切割电火花加工机床的主要部件包括床身、床鞍、工作台、立柱、臂、UV 主轴箱、切割线进给与电解系统、机床控制单元（图 12-2）。

床身：床身是较重的刚性铸件，用于支撑线切割电火花加工机床的工作部件。床身上部装有导槽（轨），对机床的主要部件起导向与调整作用。

床鞍：床鞍装在导轨上面，在进给伺服电动机与滚珠丝杠的作用下可沿 XY 方向运动。

工作台：U 形工作台装在床鞍上面，上表面均匀分布着一圈孔或螺纹孔，用于装夹工件。

立柱：立柱与床身相连，为线进给系统、UV 轴及电容开关提供支撑。

切割线进给系统：切割线进给系统为切割加工不断提供新线（电极）。新线从贮丝筒的供线端绕过一系列导轮与导柱，这些导轮和导柱给切割线提供一定张紧力。切割线沿连续路径穿过工件，用过的线又绕在贮丝筒的收线端。

工作液系统：工作液系统包括过滤器、离子交换器及冷却器。该系统可提供连续流动的恒温去离子水。去离子水能起到稳定切削状态、冲洗电极与工件上电蚀微粒以及冷却工件的作用。

机床控制单元：机床控制单元可分为三个独立控制面板。
1）设置加工条件的控制面板。
2）机床部件控制面板。
3）手动数据输入（MDI）及阴极射线管（CRT）字符显示控制面板（图12-3）。

阅读材料　　　　　　　　**线切割电火花加工（2）**

线切割电火花加工机床的各种控制系统对机床有效运行起重要作用。伺服系统、工作液、电极、机床控制单元都是线切割电火花加工机床的主要工作部件。

伺服机构

电火花加工中电源决定切削电流的大小、驱动电动机进给速度以及切割线运动速度。电火花加工机床装有伺服控制装置，可使切割线与工件之间自动保持约0.001~0.002英寸（0.025~0.05毫米）的恒定间隙。切割线（电极）与工件之间不能直接接触，这一点非常重要；否则，电弧放电会烧伤工件并熔断切割线。伺服机构还使切割线随着操作的进行而切入工件，并监测工件与切割线之间的距离，从而按需要降低或提升电动机转速，以保持适当的电弧间隙。

精确控制间隙是完成加工最基本的要求。若间隙太大，工作液不会发生电离，则加工无法进行。若间隙太小，切割线会碰上工件而熔化并烧断。

通过对比平均间隙电压与预选参考电压，电源电路可以实现精确的间隙控制。以这两个电压之间的差值作为输入信号，由它命令伺服机构切割线进给的大小与快慢，以及切割线何时从工件处回退。这通常由回退伺服灯显示出来。

当火花间隙处的切屑使电压降至临界值以下时，伺服装置使切割线回退，直到工作液冲掉切屑为止。伺服装置对冲刷切屑引起的"短暂"电压变化不应过于敏感；否则，切割线会频繁回退，严重影响加工效率。

工作液

去除加工间隙中的切屑是顺利完成电火花加工的最重要因素之一。用工作液从间隙冲出这些微粒，即可得到良好的切削环境；而冲刷不充分会使切削不稳定，并使加工条件恶化。

线切割电火花加工使用的工作液通常是去除离子的水，可由普通水循环流经离子交换树脂得到。去除离子的水是一种良好的绝缘体，未经处理的水则是导体，不适合电火花加工。离子交换剂是一种正离子交换树脂（阳离子）与负离子交换树脂（阴离子）的化合物。当水流经此化合物时，离子交换反应就开始了。该反应持续进行直到所有杂质完全从水中除去为止，这样就得到纯水。

除去离子的程度可用工作液的特性电阻衡量。对多数加工而言，电阻越小，切削速度越高。然而，当切削硬质合金材料与高密石墨材料时，所用工作液的电阻应更高一些（图12-4）。

工作液在线切割电火花加工过程中起如下几个作用：

1）有助于切割线（电极）与工件之间激发电火花。

2）在切割线与工件之间起绝缘作用。

3）冲刷切割线与工件上的切屑，以免短路。

4）对工件与切割线起冷却作用。

工作液只有在恒压下循环流动才能冲刷切屑以利改善加工状况。水流由两个控制阀控制：一个阀控制流经工件上面的水流，另一个则控制流经工件下面的水流。供水之初，先从上面提供定量小水流，然后从下面缓慢增加供水，直至工件表面骤然激起火花（图12-5）。

如果加工过程中出现红色火花，说明供水不足。解决的方法是增加供水量，直到出现蓝色火花为止。注意：过量水流会使切割线偏斜，导致切削异常，降低加工精度。

电极

线切割电火花加工所用的电极是一卷切割线，其直径在0.002～0.012英寸（0.05～0.30毫米）之间，重2～1001磅（0.09～45.36千克）。贮丝筒上切割线长度可满足500小时以上实际加工需要。线切割加工电极不断地从供线轴向收线轴移动，从而不断更新。使用此类电极，切割线的磨损不会影响加工精度，因为新线可以每分钟零点几英寸到几英寸的速度穿过工件持续进给。用作电极的切割线与工件上金属去除率均取决于如下因素：材料的导热性、材料熔点、电脉冲持续时间与强度等。与普通机床一样，电火花加工使用的部分材料比其他材料的可加工性与耐磨性更好一些。因此，电极材料应具备以下特性：

1）具有良好的导电性；

2）具有高熔点；

3）具有高张紧力；

4）具有良好的导热性；

5）可从工件上有效切削金属。

人们不断进行实验与研究，以提高金属切除率，寻求性能良好、经济实用的材料作线切割电火花加工的电极。现已发现，黄铜、铜、钨、钼、锌都是具有某些用途的电极材料。目前应用最广的电极是直径为0.008英寸（0.20毫米）的黄铜线。每边0.001英寸（0.025毫米）的过切会形成半径约0.005英寸（0.12毫米）的内圆角，因而许多加工中采用直径更小的黄铜线。钨线与钼线具有很高的熔点与张紧力，因而可以做成直径小到0.002英寸（0.05毫米）的切割线来加工半径精确而形状复杂的工件。

包层式切割线由铜线芯外包一层薄锌组成，它兼备铜的良好导电性与锌的良好散热性。这种切割线适合于大电流工作，增强了电火花加工能量，进而提高金属切除率。

Appendix 2　Vocabulary

A

ablation ［æbˈleiʃən］ *n.* 消融，切除
absolute ［ˈæbsəluːt］ *adj.* 绝对的
accommodate ［əˈkɔmədeit］ *vt.* 调节
accuracy ［ˈækjurəsi］ *n.* 精确性，准确性，精度
aerospace ［ˈɛərəˌspeis］ *n.* 航空；宇宙空间
align ［əˈlain］ *v.* 调整，校准
amperage ［ˈæmpɛəridʒ］ *n.* 安培数
anion ［ˈænaiən］ *n.* 阴离子
anode ［ˈænəud］ *n.* 极，正极
assembly ［əˈsembli］ *n.* 装配，组装
automatically ［ˌɔːtəˈmætikli］ *adv.* 自动地
automation ［ɔːtəˈmeiʃən］ *n.* 自动化；自动操作
automotive ［ɔːtəˈməutiv］ *adj.* 汽车的
axis ［ˈæksis］ *n.* 轴
axle ［ˈæksl］ *n.* 轮轴，车轴
AC servo　交流伺服系统
as long as　只要
automatic tool changer（ATC）　自动换刀装置
axis framework　坐标系

B

beam ［biːm］ *n.* 梁，桁条，横梁；（光线的）束，电波　*vt.* 播送
bedway ［ˈbedwei］ *n.* 床身导轨
bending ［ˈbendiŋ］ *n.* 弯曲
bidirectional ［ˌbaidiˈrekʃənəl］ *adj.* 双向的
bore ［bɔː, bɔː(r)］ *v.* 镗（穿、扩、钻）孔
boring ［ˈbɔːriŋ］ *n.* 镗孔
brass ［brɑːs］ *n.* 黄铜，黄铜制品；厚脸皮
breakdown ［ˈbreikˌdaun］ *n.* 崩溃；衰弱；细目分类
buffer ［ˈbʌfə］ *n.* 缓冲器
ball-nut lead screw/ball screw　滚珠丝杠
band-saw　带锯
bar stock　棒料
be referred to as　被称为
binary digit　二进制数字
binary-coded data　二进制编码数据
boring tool　钻削刀具，镗削刀具
buffer area　缓冲区
buffer storage　缓冲存储器

C

capability ［ˌkeipəˈbiliti］ *n.* 能力，性能；容量，接受力
capacitor ［kəˈpæsitə］ *n.* 电容器
carbide ［ˈkɑːbaid］ *n.* 碳化物
carriage ［ˈkæridʒ］ *n.* （机床的）滑板；刀架
cathode ［ˈkæθəud］ *n.* 阴极
cation ［ˈkætaiən］ *n.* 阳离子
chart ［tʃɑːt］ *n.* 图表
chip ［tʃip］ *n.* 碎片，碎屑　*v.* 切成碎片
chuck ［tʃʌk］ *n.* 卡盘
circuitry ［ˈsəːkitri］ *n.* 电路，线路

clamp [klæmp] v. 夹住
code [kəud] v. 编码 n. 代码，编码
coherent [kəu'hiərənt] adj. 粘在一起的，一致的，连贯的
collar ['kɔlə] n. 套环，卡圈，安装环
column ['kɔləm] n. 立柱
command [kə'mɑ:nd] n. 命令 v. 命令，指挥
compensate ['kɔmpənseit] v. 偿还，补偿
compensation [ˌkɔmpen'seiʃən] n. 补偿
component [kəm'pəunənt] n. 部件，零件
conform [kən'fɔ:m] vt. 使一致，使顺从；符合，相似，吻合
confusion [kən'fju:ʒən] n. 混乱
console [kən'səul] n. 控制台
contain [kən'tein] vt. 包含，容纳，容忍
contour ['kɔntuə] n. 轮廓
contouring [kən'tuəriŋ] n. 轮廓，造型
controller [kən'trəulə] n. 控制器
convert [kən'və:t] v. 变换，转换
coolant ['ku:lənt] n. 切削液
coordinate [kəu'ɔ:dinit] v. 调整；协调 n. 坐标
counter ['kauntə] adv. 相反地；反对地
current ['kʌrənt] n. 电流
cursor ['kə:sə] n. 光标
cast iron 铸铁
cubic boron nitride (CBN) 立方氮化硼
counter clockwise (CCW) 逆时针方向的
cemented carbide 硬质合金
circular interpolation 圆弧插补
closed-loop system 闭环系统
coded ring 编码环
computer aided programming (CAP) 计算机辅助编程
computer numerical control (CNC) 计算机数字控制
computer aided manufacturing (CAM) 计算机辅助制造
computer integrated machining (CIM) 计算机集成加工
consist of 包括
contour milling 轮廓铣削
contouring control 轮廓控制
control panel 控制面板
control unit 控制装置，控制单元
counter bore 反镗
counter boring 锪孔
cross motion 横向运动
cathode-ray tube (CRT) 阴极射线管
cutting force 切削力
cutting tool 刀具
clockwise (CW) 顺时针方向的

D

data ['deitə] n. 数据；参数
decode [di:'kəud] v. 解码，译码
deflection [di'flekʃən] n. 偏斜，偏转；偏差
deionize [di:'aiəˌnaiz] v. 除去离子
die [dai] n. 模具，冲模
dielectric [ˌdaii'lektrik] n. 电介质；绝缘体
dielectric [ˌdaii'lektrik] n. 电介质，绝缘体；a. 非传导性的
difference ['difərəns] n. 差异，差别
direct [di'rekt] v. & n. 指挥；命令
discharge [dis'tʃɑ:dʒ] vt. 放电；排出，发射 n. 放电量
dissipate ['disipeit] v. 驱散；耗散
distortion [dis'tɔ:ʃən] n. 扭曲，变形；失真
divergence [dai'və:dʒəns, di-] n. 分离，

散开；分歧

drawbar ['drɔːˌbɑː] n. 拉杆
drill [dril] n. 钻头，钻床，钻机
drilling ['driliŋ] n. 钻孔
duration [djuə'reiʃən] n. 持续时间，为期
dwell [dwel] vi. 暂停
DC servo 直流伺服系统
dimensional standard 尺寸标准
direct numerical control（DNC） 直接数字控制
distributive numerical control 分布式数字控制
drilling machine 钻床

E

effectiveness [iˈfektivnis] n. 效力
electrical [iˈlektrikəl] adj. 电的；用电的
electrode [iˈlektrəud] n. 电极；电焊条
electronic [ilekˈtrɔnik] adj. 电子的；电子器件的
emit [iˈmit] vt. 发出，放射；吐露，散发；发表，发行
encode [inˈkəud] vt. 编码，把（电文、情报等）译成电码（或密码）
encoder [inˈkəudə] n. 编码器
energize [ˈenədʒaiz] v. 提供能量
equip [iˈkwip] vt. 装备，配备；训练；准备行装
equipment [iˈkwipmənt] n. 装置，设备，装备
erode [iˈrəud] v. 腐蚀，电蚀
erosion [iˈrəuʒən] n. 腐蚀，侵蚀
erratic [iˈrætik] adj. 不稳定的
evaporate [iˈvæpəreit] vt. （使）蒸发，消失

execute [ˈeksikjuːt] vt. 执行；实行 n. 执行
expel [iksˈpel] v. 驱除，开除；排出
electrical circuit 电路
electrical device 电子设备
electrical discharge machining（EDM） 电火花加工
electrical signal 电信号
electronic element 电子元器件
end mill 面铣刀
engine lathe 卧式车床

F

fashion [ˈfæʃən] n. 方式，流行，风尚，时样
feature [ˈfiːtʃə] n. 特色，功能
feed [fiːd] n. 进给
feedback [ˈfiːdˌbæk] n. 反馈
ferrous [ˈferəs] adj. 铁的，含铁的
finish [ˈfiniʃ] n. 精加工
fixture [ˈfikstʃə] n. 夹具
flange [flændʒ] n. 凸缘，法兰
flash [flæʃ] n. 闪光，闪现，一瞬 vt. 反射 adj. 火速的
focused [ˈfəukəst] adj. 聚焦的
format [ˈfɔːmæt] n. 形式；格式 v. 对……格式化
forming [ˈfɔːmiŋ] n. （成形）加工
framework [ˈfreimwəːk] n. 构件；框架，结构
face milling 面铣削
feed rate 进给速度
feedback signal 反馈信号
feedback unit 反馈单元
figure out 合计为，计算出；解决；断定，领会到
finished product 成品

fixed-column　固定立柱

G

gage [geidʒ] v. 验，校准
generate [ˈdʒenəˌreit] vt. 产生，发生
graphite [ˈɡræfait] n. 石墨
gear cutting　齿轮加工
guide roller　导轮

H

halt [hɔːlt] n. & v. 停止，中断
harden [ˈhɑːdn] v. 使变硬；淬火
headstock [ˈhedstɔk] n. 主轴箱
horizontal [ˌhɔriˈzɔntəl] adj. 水平的
hydraulic [haiˈdrɔːlik] adj. 液压的，水压的
hydrocarbon [ˈhaidrəuˈkɑːbən] n. 烃，碳氢化合物
hard ferrous metal　硬铁合金
high-speed steel　高速钢
high-speed tape reader　高速读带机
host computer　主机
hydraulic servo　液压伺服系统

I

ideally [aiˈdiəli] adv. 理想地，在观念上地，完美地
identify [aiˈdentifai] vt. 识别，鉴别；确定
impression [imˈpreʃən] n. 印象，感想；盖印，压痕
impulse [ˈimpʌls] n. 脉冲
incorporate [inˈkɔːpəreit] v. 合并；组合
increment [ˈinkrimənt] n. 增加；增量
incremental [inkriˈmentl] adj. 增加的
individual [ˌindiˈvidjuəl] n. 个人，个体 adj. 个别的，单独的，个人的

information [ˌinfəˈmeiʃən] n. 信息
initiate [iˈniʃieit] v. 开始；发动
input [ˈinput] n. & v. 输入
inspect [inˈspekt] v. 检查
instate [inˈsteit] vt. 任命，指定
instruct [inˈstrʌkt] n. 指令 v. 命令
instruction [inˈstrʌkʃən] n. 指示，指导，指令；用法说明
integrate [ˈintiɡreit] v. 使一体化，集成
intense [inˈtens] adj. 强烈的，剧烈的
interpolation [inˌtəːpəuˈleiʃən] n. 插补
interpret [inˈtəːprit] v. 解释，说明；口译
ion [ˈaiən] n. 离子
ionization [ˌaiənaiˈzeiʃən] n. 离子化，电离
in terms of　根据，按照
indexing head　分度头
in-processing gagging　在程检测
ion exchanger　离子交换剂

J

jig [dʒig] n. 夹具

K

kerosene [ˈkerəsiːn] n. 煤油，火油
keypad [ˈkiːpæd] n. 键区
knee [niː] n. 升降台

L

lathe [leið] n. 车床 vt. 车削
lens [lenz] n. 透镜，镜头
linear [ˈliniə] adj. 线的，直线的，线性的
linear [ˈliniə] adj. 直线的，线性的
load [ləud] vt. & vi. 加载
logic [ˈlɔdʒik] n. 逻辑；逻辑性
laser cutting machine　激光切割机床

lead screw 丝杠
lengthwise travel 纵向运动
line motion control 直线运动控制
linear interpolation 直线插补
linear movement/motion 直线运动
loop system 回路系统

M

machining [məˈʃiːniŋ] v. 机械加工
mainframe [ˈmeinfreim] n. 主机，大型机
manually [ˈmænjuəli] adv. 手动地，人工地
manufacture [ˌmænjuˈfæktʃə] v. 制造，加工 n. 制造，制造业；产品
medium [ˈmiːdjəm] n. 介质，媒介；方法
metalworking [ˈmetəlˌwəːkiŋ] n. 金属加工术
micro [ˈmaikrəu] adj. 微小的
microelectronics [ˌmaikrəuiˌlekˈtrɔniks] n. 微电子学
mill [mil] n. 铣床；铣刀
milling [ˈmiliŋ] n. 铣削
molybdenum [məˈlibdinəm] n. 钼
monochromatic [ˌmɔnəukrəuˈmætik] adj. 单色的，单频的
motion [ˈməuʃən] n. & v. 运动
mount [maunt] n. 装备 v. 安装；设置；安放
multifunction [ˌmʌltiˈfʌŋkʃən] n. 多功能
machine control unit (MCU) 机床控制单元
machine tool 机床
machining center 加工中心
machining process 加工过程
manual data input (MDI) 手动数据输入
manual programming 手工编程
machine control unit (MCU) 机床控制单元
manual data input (MDI) 手动数据输入
metal cutting 金属切削
milling cutter 铣刀
miscellaneous/auxiliary function 辅助功能
mold making 模具制作

N

network [ˈnetwəːk] n. 网络
nonconductor [ˈnɔnkənˈdʌktə] n. 绝缘体
nontraditional [ˌnɔntrəˈdiʃənəl] adj. 非传统的，不符合传统的
numerical [njuː(ː)ˈmerikəl] adj. 数字的
non-chip-producing time 非切削时间
nonferrous material 非铁材料
numerical control (NC) 数字控制（数控）

O

onwards [ˈɔnwəːdz] adv. 向前地，在先地
operate [ˈɔpəreit] v. 运转；操作
operator [ˈɔpəreitə] n. （机器、设备等的）操作员
order [ˈɔːdə] n. 次序，顺序；命令；定购 vt. 定购，定制
orthogonal [ɔːˈθɔɡənl] adj. 直角的，正交的
output [ˈautput] n. 输出
on-board 在船（飞机，车）上，文中指内嵌在机床上
open-loop system 开环系统
orthogonal axis frame 直角坐标系

P

pallet ['pælit] *n.* 托盘，平板架
panel ['pænəl] *n.* 面板
part [pɑːt] *n.* 零件，工件
pause [pɔːz] *n.* & *v.* 暂停
penetrate ['penitreit] *vt.* & *vi.* 穿过，进入；了解
plural ['pluərəl] *adj.* 复数的
pneumatic [njuː'mætik] *adj.* 风力的；气压的
pocketing ['pɔkitiŋ] *n.* 凹槽
polycrystalline [ˌpɔli'kristəlain] *adj.* 多晶的
portion ['pɔːʃən] *n.* 一部分，一份
positional [pə'ziʃənl] *adj.* 定位的，位置的
precision [pri'siʒən] *n.* 精度
predetermine ['priːdi'təːmin] *v.* 预定，预先确定
preprogram [priː'prəugræm] *v.* 预编程序
process [prə'ses] *n.* 过程，步骤 *v.* 加工，处理
profile ['prəufail] *n.* 剖面，侧面；外形，轮廓
program ['prəugræm] *vt.* （为……）编（制）程序 *n.* 程序
pulse [pʌls] *n.* 脉冲
punch [pʌntʃ] *v.* 冲孔，打孔 *n.* 冲压机，冲床
peripheral equipment 外围设备
personal computer (PC) 个人计算机
pneumatic device 气动设备
point-to-point control 点位控制
production rate 生产率
program reference zero (PRZ) 程序参考零点

prototype work 标准工件
punched card 穿孔卡
punched tape 穿孔带

Q

quill [kwil] *n.* 衬套；主轴
quadratic arcs 二次曲线

R

radiation [ˌreidi'eiʃən] *n.* 发散，发光，发热，辐射；放射线，放射物
ream [riːm] *v.* 铰孔
reamer ['riːmə] *n.* 铰刀，扩锥
reaming [riːmiŋ] *n.* 铰孔
recall [ri'kɔːl] *v.* & *n.* 调用
receive [ri'siːv] *v.* 接收
rectangular [rek'tæŋgjulə] *adj.* 长方形的
reference ['refrəns] *n.* 参考
register ['redʒistə] *n.* 寄存器
regrind [riː'graind] *v.* 重磨
relay [ri'lei] *v.* 传送 *n.* 继电器
remove [ri'muːv] *vt.* 拿走，去掉
represent [ˌriːpri'zent] *vt.* 代表；声称；表现
resin ['rezin] *n.* 树脂
retract [ri'trækt] *v.* 回退，缩回，缩进，所卷进（舌等）；收回，取消，撤销
retrofit ['retrəˌfit] *n.* 改型（装，进）；（式样）翻新
revolver [ri'vɔlvə] *n.* 旋转变压器
rigidity [ri'dʒiditi] *n.* 刚度
rod [rɔd] *n.* 杆，棒
roller ['rəulə] *n.* 滚筒，辊子，导轮（用在线切割机床上）
rotary ['rəutəri] *adj.* 旋转的
rough [rʌf] *n.* 粗加工
rugged ['rʌgid] *adj.* 结实的，坚固的

random-access memory（RAM） 随机存储器
reference plane R 点平面，参考平面
right-hand rule 右手法则
rotary movement/motion 旋转运动
rough casting 铸造毛坯

S

saddle ['sædl] n. 鞍，鞍状物
saving ['seiviŋ] n. 节约；存款；挽救，救助
schematically [ski'mætikli] adv. 示意性地
scheme [ski:m] n. 配置，图解；计划，设计，图谋，策划
semiautomatic ['semi,ɔ:tə'mætik] adj. 半自动的
send [send] v. 发送
sense [sens] n. 感觉，判断力，理性感；理解，认识
sequence ['si:kwəns] n. 次序，顺序，序列
sequential [si'kwinʃəl] adj. 连续的；有序的；结果的
servo ['sə:vəu] n. 伺服系统
servocontrol ['sə:vəkən'trəul] n. 伺服控制，随动控制
servomechanism ['sə:vəu'mekənizəm] n. 伺服机构（系统），自动控制装置
servomotor ['sə:vəu'məutə] n. 伺服电动机
setup ['setʌp] n. 安装；设备；机构
sheet [ʃi:t] n. （一）片，（一）张；清单
shop [ʃɔp] n. 商店；工厂，修理厂
shuttle ['ʃʌtl] n. 滑闸，滑台 v. 穿梭往返

signal ['signəl] n. 信号 v. 发信号
simplicity [sim'plisiti] n. 简单，朴素，直率
simultaneously [,siməl'teiniəsli] adv. 同时发生地，同步地
slide [slaid] n. 滑板
slot [slɔt] n. 狭缝，窄槽
software ['sɔftwɛə] n. 软件
solenoid ['səulinɔid] n. 螺线管
spark [spɑ:k] n. 火花，火星，电火花
specify ['spesifai] vt. 指定；详细说明，列入清单
spindle ['spindl] n. 轴，杆，心轴
spool [spu:l] n. 线轴
stamping ['stæmpiŋ] n. 冲压
stem [stem] vi. 起源
stepper ['stepə] n. 步进者
storage ['stɔridʒ] n. 贮藏，存储
store [stɔ:] v. 存储，储藏 n. 储备
stratified ['strætifaid] adj. 分层的
strength [streŋθ] n. 力；强度
stretch [stretʃ] v. 伸出；拉紧
stud [stʌd] n. 双头螺栓；销子；拉钉
submerge [səb'mə:dʒ] vt. 浸没，淹没，湮没 vi. 潜水
substantial [səb'stænʃəl] adj. 坚固的；实质的，真实的；充实的
subtraction [səb'trækʃən] n. 减少
symmetrical [si'metrikəl] adj. 对称的，均匀的
synthetic [sin'θetic] adj. 合成的
servo control 伺服控制
servo control unit 伺服控制单元
shell end mill 圆筒形面铣刀（或套筒铣刀）
spark gap 火花隙
spot-facing 刮孔口平面的

stepping motor　步进电动机

T

table ['teibl] *n.* 工作台
tap [tæp] *n.* 丝锥
taper ['teipə] *n.* 锥度
tapping ['tæpiŋ] *n.* 攻螺纹
technology [tek'nɔlədʒi] *n.* 工艺，技术
temporary ['tempərəri] *adj.* 暂时的，临时的，临时性
thermoelectric [ˌθəːməui'lektrik] *adj.* 热电的
thickness ['θiknis] *n.* 厚度，浓度；稠密
thread [θred] *n.* （螺纹的）头数
toolholder ['tuːlˌhəuldə(r)] *n.* 刀柄
tooling ['tuːliŋ] *n.* 加工；刀具，工具
torque [tɔːk] *n.* 转矩
transducer [træns'djuːsə] *n.* 传感器；变频器
transfer [træns'fəː] *v.* 传递；改变　*n.* 传递；转移
translate [træns'leit] *v.* 转变为；翻译
trigger ['trigə] *vt.* 引发，引起，触发　*n.* 扳机
tungsten ['tʌŋstən] *n.* 钨
turning ['təːniŋ] *n.* 车削
turret ['təːrit] *n.* 转塔刀架
take over　接受，接管
tape reader　读带机
tool changer　换刀机构
tool changing (tool change)　换刀
tool code　刀具代码
tool length offset　刀具长度偏置
tool mark　刀痕
tool nose radius compensation　刀尖圆弧半径补偿
tool radius compensation　刀具半径补偿
tooling cost　刀具加工成本
tooling industry　刀具业
tool-storage　刀具存放
torsional strength　扭转强度
traveling-column　移动立柱
turn off　关闭
turn on　打开
turning center　车削中心
turret head　转塔头

U

uncompensated [ʌn'kɔmpenseitid] *adj.* 未得补偿的，没有得到赔偿的
undergo [ˌʌndə'gəu] *vt.* 经历，遭受，忍受
unload [ˌʌn'ləud] *vi.* 卸载

V

vaporize ['veipəraiz] *v.* 蒸发，汽化
verify ['verifai] *vt.* 检验，校验，查证，核实
vertical ['vəːtikəl] *adj.* 垂直的，直立的
visualize ['viʒjuəlaiz] *vt.* 可视化，形象化；想象　*vi.* 显现
vacuum tube　真空管
vertical milling machine　立式铣床

W

wavelength ['weivˌleŋθ] *n.* 波长
welding ['weldiŋ] *n.* 焊接
wire ['waiə] *n.* 电线；电报，电信；金属丝，铁丝网　*vt.* 用金属线捆扎、联结或加固
workpiece ['wəːkˌpiːs] *n.* 工件
wire-cut electrical discharge machining (wire-cut EDM)　线切割电火花加工机床

work piece 工件
work-holding 工件夹紧

X

xenon [ˈzenɔn] n. 氙

Z

zinc [ziŋk] n. 锌
zero preset 零点预置

References

[1] 王兆奇. 数控专业英语 [M]. 北京：机械工业出版社，2007.

[2] 章跃，张国生. 机械制造专业英语 [M]. 北京：机械工业出版社，2001.

[3] 沈言锦，周钢. 模具专业英语 [M]. 北京：机械工业出版社，2009.

[4] 戴浩中. 机械英语自学读本 [M]. 上海：上海科学技术出版社，1983.

[5] 刘瑛，王莉. 数控技术专业英语 [M]. 2 版. 北京：人民邮电出版社，2008.

[6] 施平. 机械工程专业英语 [M]. 哈尔滨：哈尔滨工业大学出版社，1998.

[7] 鲍海龙. 数控专业英语 [M]. 北京：机械工业出版社，2007.

[8] 刘振康. 机械制造英语读本 [M]. 北京：机械工业出版社，1988.

[9] Steve F Krar, Albert F Check. Technology of Machine Tools [M]. 5th ed. New York：Glencoe/McGraw Hill，1997.

[10] Jon Stenerson, Kelly Curran. Computer Numerical Control [M]. New Jersey：Prentice Hall，1997.